AF604279

WORLD'S MOST INCREDIBLE ANIMALS

Words by Liz Ginis,
Karen McGhee
and Pete Tuskan

CONTENTS

Eastern gorillas are gentle giants – and sadly they are critically endangered due to habitat destruction.

African lions can weigh as much as 190kg, run at 80km/h, and are currently listed as vulnerable.

DEADLY

These are the animals that you only want to see from a distance. Be extra careful as we take you through the most DEADLY creatures on our planet!

BLACK MAMBA

This snake is one of the most feared animals in Africa, with a reputation for delivering the "kiss of death". No one has been known to survive its highly toxic venom without treatment, and each year it kills thousands of people across the continent. That's partly because the black mamba is so widespread in Africa, and people often come across it in remote areas, so can't reach treatment. It's also extremely aggressive.

DEADLY FACT!

The black mamba is the fastest land snake in the world, and is able to move faster than most humans can run.

PUFF ADDER

No other snake causes as many deaths in Africa as this species. It occurs from about the middle of the continent down, and is often found living around villages and towns. It's an ambush predator that relies on camouflage to keep hidden while it waits for prey to pass by – then it strikes with lightning speed. Usually, however, it's a slow-moving, cranky reptile, and most of its victims stumble across it by accident. The good news is that its venom is slow-acting, and with treatment most victims can be saved.

EGYPTIAN COBRA

This species was worshipped by the ancient Egyptians. Its bite can kill a human in just 15 minutes, and bring down an adult elephant in three hours. The Egyptian cobra can be found living across northern Africa. Known also as the asp, it's the cobra species famed for its use in shows by African snake charmers. Despite being one of the continent's most deadly snakes, this cobra is quite docile, and will sometimes even play dead when faced with a predator, such as a mongoose.

PHARAOH DEATH

It's thought that this is the species of snake whose venom Cleopatra, the last Egyptian leader of the Ptolemaic Dynasty, died from in 30 BC.

BLACK RHINOCEROS

Black rhinos weigh up to 1.5 tonnes (although a few have hit 3 tonnes!), and have large horns at the front of their snouts. Combine these features with an ability to reach a speed of more than 60km/h if they decide to charge, and you'll want to keep well out of their way. On top of all this they've got poor eyesight, a great sense of smell, and an unpredictable nature that makes them even more dangerous.

SPOTTED HYENA

Hyenas have a reputation as scavengers who feed only on carrion, but they are among Africa's very best hunters. They work together in packs, which can sometimes include as many as 70 animals. They're very good at chasing down large prey over distances, during which they can run for long periods of time at up to 60km/h. Fortunately, if you keep your distance, they'll usually leave you alone.

PLAYMATES?

Lions and hyenas often share territories, as they live in similar habitats, and will regularly squabble over fresh meat.

DEATHSTALKER SCORPION

This deadly creature is found in the desert and dry scrubland of north-eastern Africa, and has the most potent venom of any scorpion. An antivenom is available, but those who don't get it fast enough will unfortunately often die. Deathstalkers are large as far as scorpions go, up to 11cm long. To avoid the heat, they live in burrows by day and only emerge at night.

NEW USES

A substance called chlorotoxin, found in deathstalker venom, has been used to develop a new treatment for people with brain cancer.

CAPE BUFFALO

The Cape buffalo is a grazing herbivore, but it's surprisingly aggressive, and is responsible for the deaths of around 200 hunters in Africa every year, where it's known as Black Death. They appear to be able to recognise and remember hunters who have previously tried to cause them or their herd harm, and have been known to ambush them and wound them when they return, even after many years.

WARNING SIGN

Cape buffalo will kill the cubs of adult lions that have attacked their herd, as a warning to not do it again. A very clever preventative punishment.

HIPPOPOTAMUS

They might be vegetarians, but hippos are actually highly aggressive creatures, and cause more human deaths in Africa than any other large animal – according to reports, they kill up to 500 people every year. The males are extremely territorial, and will readily throw their weight, which can be as much as 2 tonnes, around to keep everyone and everything away from their patch of riverbank. They also have huge jaws containing canine teeth that can be up to 50cm long! And female hippos are just as dangerous, because they're extremely protective of their young.

TSETSE FLY

This blood-sucking insect looks very much like an ordinary house fly, but it's responsible for the debilitating and often fatal disease known as sleeping sickness, of which there are around 20,000 new cases in Africa every year. Victims inevitably die if they don't receive treatment. The disease is caused by a tiny single-celled parasite that's passed on by the tsetse fly, which it picks up from a person or animal that's already infected.

NILE CROCODILE

You don't ever want to get close to the jaws of a Nile crocodile. They're weapons that bite down with a force more powerful than any other animal, and they are studded with dozens of razor-sharp teeth too. If this animal got you in its mouth and the bite didn't kill you, you'd drown anyway because they take all their prey underwater soon after they grab it. Every year, crocodiles claim the lives of hundreds of people in Africa. These victims have usually made the mistake of wandering too close to the edges of rivers or lakes where crocodiles lurk motionless beneath the water.

WOLVERINE

Although they look like bears, only grow as big as medium-sized dogs, and are relatives of weasels, there are good reasons why this creature inspired the aggressive comic book and movie character. They are solitary hunters found across the northern parts of Europe, Asia and North America, near Arctic areas, and what makes them such good hunters – and why we should stay clear of them – is that they are persistent, ferocious and incredibly strong for their size. They also have sharp claws and powerful jaws that can chomp through bone with ease. Wolverines eat a lot of smaller prey such as rabbits, but they'll also attack and bring down animals that are a lot bigger than them, such as reindeer.

WOLVERINES HAVE BEEN KNOWN TO STOCKPILE THEIR FOOD FOR THE FUTURE. THEY USE THE SNOW LIKE A REFRIGERATOR, KEEPING THEIR FOOD FRESH FOR TIMES WHEN THEY AREN'T ABLE TO HUNT.

ASP VIPER

Europe has fewer venomous snakes than any other continent bar Antarctica, but there are a few species capable of killing people, and the asp viper is the most dangerous of them. Like other vipers, it has large fangs that fold back when the mouth is closed. These are hollow, and work like hypodermic syringes to inject venom into its victims, which makes the bite of the asp viper particularly painful. Despite the pain, most humans will survive being bitten, however every year a small percentage of people have died because they couldn't receive medical treatment in time. Asp vipers don't seek out humans though – small mammals are their main prey, and they'll also take lizards or birds.

ASP VIPERS ARE USUALLY SLOW-MOVING AND TIMID, BUT CORNER THEM AND YOU'LL REALISE THAT THEY CAN BE VERY AGGRESSIVE, AND CAPABLE OF STRIKING OUT.

EURASIAN WOLF

This is the big bad wolf you hear about in fairytales, and there's much evidence from the past to show that these wolves deserved their bad reputation. Historically, many people have been killed by wolves in Europe, particularly in France and Eastern Europe, and this is the continent's largest species. But it now seems many of these attacks involved animals infected with rabies, a virus that affects the central nervous system. Wolves with the disease lose their fear of humans and become extremely aggressive. However, rabies is not nearly as common in Europe today as it once was, and wolf attacks on humans are now extremely rare.

DEAR OH DEER

Wolves are extremely intelligent, and in the wild they live and hunt together in packs to bring down animals that are much larger than them, such as deer.

POLAR BEAR

It doesn't take much for a polar bear to want to kill you. Just cross their path, and if they're hungry – which they usually are – you're not likely to survive. This is the world's largest land carnivore, and it has no natural fear of people. Males reach an average weight of about 400kg, and are the length of a family car – when they stand on their hind legs they can be more than 3m tall. Don't think that being big makes them slow either – they can run faster than Olympic sprinters, at 40km/h. Because these bears usually live in isolated areas, humans don't often encounter them, but with their habitat now shrinking in size due to climate change, polar bears are increasingly wandering into towns around the Arctic Circle looking for food.

MEDITERRANEAN BLACK WIDOW SPIDER

This is Europe's most venomous spider. Its venom contains a potent nerve poison called latrotoxin. Symptoms after being bitten include sweating and muscle spasms. Victims can experience difficulty breathing, develop an irregular heart rate, vomit – or even die. Farm workers in the European grasslands are bitten most often.

ONLY THE BITE OF THE FEMALE OF THE MEDITERRANEAN BLACK WIDOW SPIDER SPECIES IS DANGEROUS TO HUMANS.

PORTUGUESE MAN O' WAR

The tentacles of the Portuguese man o'war are covered in stinging cells called nematocysts. These contain venom that stuns and kills prey that passes too close. Unfortunately, people swimming along Europe's coast during spring and summer are often stung. Occasionally the stings cause life-threatening allergic reactions – or they can be so painful that they cause shock that leads to drowning.

Known in Australia as bluebottles, these floating creatures are blown all around the world, from Europe to the USA and down to the Southern Hemisphere.

ALTHOUGH REFERRED TO AS A KILLER WHALE – DUE TO ITS SKILL AT HUNTING WHALES – THE ORCA IS ACTUALLY A DOLPHIN.

ORCA

To whales, these supreme hunters of the sea are as fearsome as sharks, and maybe even a little more formidable due to their extraordinary intelligence. Orcas have mouths full of 8cm-long teeth, and big brains that enable them to communicate with each other so they can hunt together in packs. The largest species of dolphin, orcas also have size and speed on their side, growing to more than five tonnes, and able to reach speeds of 56km/h. They live in all oceans, from polar regions to the equator, but are most abundant in colder waters such as northern Europe. Efficient hunters, the common name killer whale is fitting for orcas. Humans don't tend to swim in the same waters they do though, so aren't usually threatened by them, but a few captive orcas have harmed or killed their human handlers. Rather, these cetaceans prefer to chase their usual prey of fish, seals, and the occasional whale.

WILD BOAR

It's their tusks that make wild boars so dangerous. These are their incisor teeth, which grow particularly large and sharpen as they mature, with males having larger sets than females. Mix these weapons with the aggressive territorial behaviour that many wild boars display, and you've got creatures that can rip you open if they think you're threatening them. Back off very quickly if you come across one of these animals in the European countryside.

OINK

Domesticated pigs were originally bred from wild boars. Today, wild boars also interbreed with farmyard pigs that have escaped into the wild.

ASIA

BENGAL TIGER

Most lists of the world's deadliest animals have the Bengal tiger close to the top. These big cats are thought to have killed more than 370,000 people in Asia during the 19th and 20th centuries. Bengals are the second largest big cats living today, and one of the five surviving tiger sub-species. Males can grow up to 3m long and weigh 300kg. They're big and powerful, and also very territorial. They live and hunt in areas where humans have set up villages, near grasslands at the edge of forests, which means these tigers are more likely to come into contact with people than most other big cats. Tigers usually prefer prey such as buffalo, wild boar and deer, but there are reports of rogue tigers that have become man-eaters. Among the most famous was the so-called Tigress of Champawat, who lived in Nepal during the 19th and early 20th centuries. After being shot by a hunter she was unable to go after her usual prey, and turned to humans. She's said to have killed more than 400 people before eventually being shot dead.

DEADLY FACT!

Tigers don't usually kill by ripping their prey apart. Instead, they will lock their powerful jaws around their prey's neck and suffocate it to death.

TIGER SHARK

More than 450 species of shark have been identified, but only six are known to attack humans. Of these, tiger sharks are one of the worst offenders. According to the International Shark Attack File, 34 people have been killed by tiger sharks in the past few hundred years, and 97 have survived attacks.

Found in tropical coastal waters throughout Asia, as well as in many other parts of the world, their jaws are so powerful and their serrated teeth so sharp that they can crack through the hard shells of sea turtles.

DEADLY FACT!

When tiger sharks are young they have vertical stripes on their body, which is where their common name comes from. These mostly disappear as they mature.

SLOTH BEAR

Sloth bears might primarily be insect-eaters, but they react to predators with extreme aggression. They will attack anyone in their territory, particularly if they think their cubs are threatened. Sloth bears have long claws to help them break into termite mounds, and they use these when attacking humans, typically striking at the face and head. Sloth bears live mostly alone in the forests of Sri Lanka, India, Bangladesh and Bhutan.

AS WELL AS INSECTS, SLOTH BEARS EAT FRUITS AND FLOWERS, AND OCCASIONALLY BREAK INTO BEE HIVES FOR THE HONEY.

FAT-TAILED SCORPION

This one has the most lethal venom of all scorpions. These eight-legged creatures are only about 10cm long, but the venom in their tail is such a powerful nerve poison that it can kill in less than an hour. Each year, a few unlucky people are killed by fat-tailed scorpions – whose scientific name means man killer – usually in their natural habitat, the hot dry regions of India, northern Africa and the Middle East.

QUICK SMART!

A scorpion antivenom is available. If you're ever stung, get to a hospital as fast as possible for a life-saving shot.

KING COBRA

A bite from this slithering reptile can bring down an elephant or kill a person within 30 minutes. Luckily, king cobras are just as scared of people as we are of them, and would prefer to use scare tactics before taking a bite. When threatened, they'll rise up to the height of an adult human, with about a third of their body off the ground. Then they'll flatten and flare out the skin around their head like a hood, hiss loudly, and lunge forward as if they are going to strike. The king cobra's venom isn't quite as toxic as that of other deadly snakes, but it makes up for that by producing huge quantities of it. It's a mixture of deadly chemicals, but the main one is a powerful nerve poison called haditoxin. Victims experience excruciating pain, then become sleepy, dizzy and paralysed. The effects can be reversed with antivenom, often the same type developed to treat the bites of Australian tiger snakes. King cobras are found mostly in the forests of India, China, Malaysia, Indonesia and the Philippines.

THE KING COBRA IS THE WORLD'S LARGEST VENOMOUS SNAKE. IT CAN GROW TO A LENGTH OF 5.6M.

ASIAN GIANT HORNET

These insects are often referred to as murder hornets because they are so lethal. In Japan, they kill up to 50 people a year. With bodies up to 5.5cm in length and an impressive 7cm wingspan, they are the largest hornets on earth. Their stinger is more than 6cm long, and their stings are incredibly painful – like hot metal driving into the skin, according to victims – and can cause an anaphylactic response. They have also been recorded spraying their venom into people's eyes. Thankfully, they are only aggressive when they feel threatened or are defending their nest.

These hornets like eating tree sap and honey, but their main source of food is other insects, including beetles and honey bees. This species of hornet can kill up to 40 bees in a minute, and entire colonies have been devastated by hornet attacks. Once a hornet has its prey in hand, they will decapitate their victim before chewing them into a paste. The Asian giant hornet has a big orange head and striped body, making them easy to identify in the wild. They are native to temperate and tropical Asian coastlines, but there have been recent sightings in the US too.

DEADLY FACT!

The stinger on an Asian giant hornet is more than 6cm in length, so you definitely don't want to upset them!

MOSQUITO

These tiny insects are big killers – the world's biggest, in fact! Mosquitoes kill more people than any other animal does; about one million a year. What makes them so deadly are the infections they transfer into the bloodstreams of their victims. Mosquitoes thrive in the warm, wet conditions found throughout tropical Asia, where the main killer disease they carry is malaria, as well as dengue fever, yellow fever, encephalitis, zika and chikungunya. They're not limited to this region however – there are 400,000 deaths a year in Africa from malaria alone, and in Australia mozzies transmit potentially chronic illnesses such as Ross River fever and Murray Valley encephalitis.

DID YOU KNOW?

It's only female mosquitoes that bite – they need the protein in blood to produce eggs. Male mosquitoes feed on nectar, like butterflies do.

ASIATIC LION

The Asiatic lion is a subspecies of lion that's slightly smaller than its better-known African cousin. A formidable predator, it hunts in family groups, and has been known to kill humans. However, the Asiatic lion is now so at risk that humans are more of a threat to it than the other way around. The species used to live throughout Asia, but it was hunted so intensively that by the early 1900s, there were only 12 left. Classified as endangered and now protected, today there are about 650 lions.

LAST STAND

The remaining lions live in a sanctuary in Gir Forest in India, where the population is slowly increasing.

BANDED KRAIT SNAKE

These stunningly marked snakes are as dangerous as they are beautiful. Their venom is extremely toxic and potentially fatal to humans. The good news is that banded kraits aren't aggressive – in fact they're so docile that they're unlikely to attack a person, even when they're being hassled or provoked. But if they *do* bite, the severe injection of venom can lead to respiratory failure.

The banded krait is known to feed mainly on other snakes, biting them before swallowing them head first.

KOMODO DRAGON

Male Komodo dragons can grow to a length of more than 3m and a weight of about 160kg, and they've got the jaws, teeth and claws you'd expect to see on a voracious meat-eater of this size. These massive reptiles are the world's biggest lizard. They usually prey on buffalo, wild pig and deer, but have been known to occasionally attack humans. Fortunately people don't encounter them very often, as they're only found on Indonesia's remote Komodo Island and some of the smaller surrounding islands. The Komodo is an ambush hunter that launches surprise attacks on animals that pass too close, making short, fast lunges while grabbing with its powerful claws and serrated shark-like teeth. In 2009, it was discovered that Komodo dragons also produce venom. This stops mammal blood from clotting, so even if prey manages to get away, it will ultimately bleed to death. The Komodo will track it patiently, waiting until this happens, then move in for its meal.

DEADLY FACT!

Komodo dragons hear and see well, and have an exceptional sense of smell. They pick up odours by flicking out their forked tongues to "taste" the air, like snakes do.

OCEANIA

SALTWATER CROCODILE

Every year up to 10 people are attacked in Australia by saltwater crocodiles, with one or two of them being fatal. Salties live in northern Australia, in freshwater river systems, wetlands and along the coast. Their favourite attack zones are mangrove swamps and riverbanks, and they can even be spotted swimming a long way out at sea. They breathe air, but have special adaptations that mean they can remain submerged for at least an hour, lurking silently just below the surface until they explode out and grab prey that comes too close. Once caught, they'll roll back in the water with it in a move known as a death roll. This either breaks their prey's neck or drowns it. If their prey is large, they'll tear it into pieces and bury it in a kind of larder that they can come back to when they're hungry.

DEADLY FACT!

This is the world's biggest living reptile. Males can grow longer than 7m and weigh a tonne – about the size of a family car.

HOODED PITOHUI

Native to New Guinea, this bird is both beautiful and dangerous. It is one of three pitohui species – all are poisonous, with the hooded one being the most deadly. These pretty songbirds have a neurotoxin in their skin and feathers called homobatrachotoxin, which causes numbing on contact. They acquire their poison from eating the choresine beetle.

LOCALS RUB CHARCOAL ONTO THE BIRDS BEFORE COOKING TO NULLIFY THE POISON'S EFFECTS.

BULL SHARK

This aggressive shark has powerful jaws and eats almost anything. It's common in warm, shallow coastal waters, and is the only shark that can handle fresh water for long periods. This means it can leave the ocean and swim up river systems, where it's been known to take cattle, dogs, and occasionally people. It also comes into the shallows on a rising ocean tide looking for anything edible that may have floated off the shore.

SYDNEY FUNNEL-WEB

Sydney funnel-webs are among the scariest looking spiders, so it's fitting that they're also one of the deadliest. Females have hairy black bodies up to 5cm long, with the more toxic males slightly smaller. They put on a very fierce display when threatened, rearing up, lifting the body and front legs, revealing their fangs, then thrusting them down frighteningly fast. The fangs are only about 6mm long, but they are capable of stabbing through a toenail! Even discounting the venom, these fangs hurt when they puncture skin! Fortunately there have been no deaths recorded in Australia since the introduction of an antivenom in 1981.

BOX JELLYFISH

There are lots of different box jellyfish around the world, and the marine creature with this name in Australia is one of the deadliest animals on the planet. Its sting can kill a healthy adult in minutes. Box jellyfish venom contains toxins that attack the nerves, blood and organs. It also contains dermatonecrotic substances, which scar the skin. For that reason anyone lucky enough to survive an encounter with this creature will have lifelong scars. This jellyfish has as many as 60 tentacles, each up to 3m long. Venom is stored along these in millions of microscopic stinging cells called nematocysts, which each contain a tiny harpoon that bursts out on contact to inject venom. Your chance of dying from a box jellyfish sting depends on the length of tentacles you come in contact with, and getting to treatment in time.

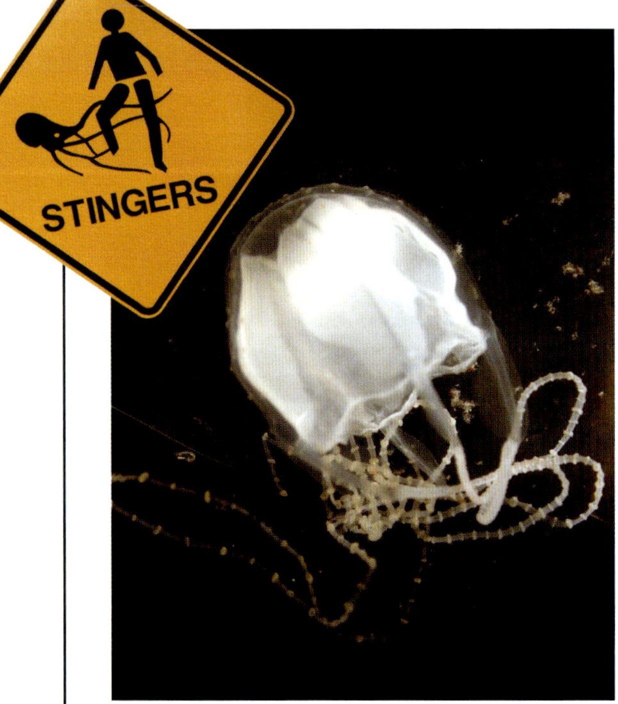

IRUKANDJI JELLYFISH

Irukandji is a venomous jellyfish found in northern Australian waters during summer. You don't always notice when you're being stung, but the full effect will soon hit you. Irukandji syndrome includes severe lower back pain and full body muscle cramps, along with sweating, anxiety, restlessness, nausea and vomiting. Headaches and heart palpitations develop, along with life-threatening high blood pressure and fluid build-up, and in extreme cases a sting may result in heart failure, swelling of the brain and death.

Most Irukandji have bodies that are no wider than 3cm, and thin wispy tentacles up to a metre long.

EASTERN BROWN SNAKE

This is Australia's most deadly snake. No other snake bites or kills more people in Australia than the eastern brown. It's up to 2m long and has small fangs, but its venom packs more of a punch than most other snakes. It's ranked as the second-most toxic of any land snake in the world, after the inland taipan, and has caused the deaths of at least 24 people since 1980. What makes this snake extra dangerous is its behaviour. It's fast-moving, aggressive, and active during the day, and is found along the east coast, where most of Australia's population lives. All of this means that it's also the snake most Aussies are likely to come in contact with, increasing the chances of being bitten.

EASTERN BROWN SNAKES MAINLY EAT MAMMALS, FROGS, BIRDS, REPTILES AND EGGS.

BLUE-RINGED OCTOPUS

The world's most lethal octopus – the venom in one sting is powerful enough to kill 26 adults – lurks in crevices in rock pools along the coast. They're responsible for two known deaths in Australia; fortunately most people survive. Fast-acting venom can cause muscle weakness and paralysis within 10 minutes, and death can take just 30 minutes, as a result of the lack of oxygen to the brain due to breathing problems.

SANS VENOM

There's no antivenom available, but victims can survive if they are given CPR and placed on a respirator.

COASTAL TAIPAN

Before the introduction of an antivenom in 1956, taipan bites caused many deaths in Australia. Their venom affects the blood and nervous system of victims, and causes nausea, internal bleeding, muscle destruction and kidney damage. In severe cases, death can occur in just 30 minutes. It's a nervous snake that does its best to stay out of the way of people, but if cornered it can be ferocious.

Coastal taipans have evolved so that their venom works especially well on warm-blooded animals.

GRIZZLY BEAR

All types of North American brown bears are called grizzlies. They're carnivores that will take large prey such as moose, elk and caribou if they can, but they also eat a lot of fish, fruits and berries. Each year, there's at least one or two fatal attacks on humans by grizzlies. These usually occur when people have suddenly surprised a bear while walking in its territory. For this reason, making a lot of noise when you're in bear country is recommended, so they know you're there. All encounters with grizzlies are potentially deadly, but running away once a bear is committed to an attack isn't an option, because they can run at 60km/h. Mothers with cubs can be particularly aggressive and dangerous.

DEADLY FACT!

Adult males can weigh more than 380kg, and be up to 3m tall when they're on their hind feet. Grizzly bears usually walk on all fours, but can rise up on two. It's usually a display of curiosity rather than aggression, but you don't want to test it out!

AMERICAN ALLIGATOR

By the late 1960s in North America, these alligators were dangerously close to extinction, due to hunting and the destruction of their wetlands habitat, but after being placed on the endangered species list, their numbers are now back up and their survival seems assured. These carnivorous reptiles can reach huge proportions – they're the second-largest species in the alligatoridae family, with adult males growing up to a huge 4.5m long and weighing about half a tonne. They are apex predators, consuming a diet of amphibians, fish, other reptiles, mammals and birds, and are less aggressive towards humans than crocodiles are. However, there has been an increase in attacks over the past 20 years, with 22 people confirmed killed by American alligators in the south-west of the United States in the past two decades.

AMERICAN ALLIGATORS CAN LIVE FOR UP TO 50 YEARS IN THE WILD. ONE OF THE REASONS THEY LIVE SO LONG IS THAT ONCE THEY REACH A LENGTH OF AROUND A METRE, THEY HAVE NO PREDATORS, EXCEPT FOR HUMANS.

EASTERN DIAMONDBACK RATTLESNAKE

This is the largest venomous snake in the US, weighing up to 4.5kg, and up to 2.4m long. There are five deaths a year from snake bite in the US, with rattlesnakes being the worst offenders, however they prefer to keep their venom for prey and not waste it. That's what the rattles on the ends of their tails are for – to warn potential predators or large grazing animals that might step on them to stay away. Rattlesnake venom contains a potent mix of toxins that attack the nervous system as well as blood and other tissues, but there is now an antivenom that saves thousands of lives every year.

BLACK WIDOW SPIDER

The bite of a black widow spider is rarely fatal, but it can make victims very sick, with symptoms including aches, chills and fever, nausea and vomiting, severe stomach pains and extremely high blood pressure. Only female black widows bite, and only when they're disturbed while protecting their eggs.

EYE SPY

There's no antivenom available, but victims can survive if they are given CPR and placed on a respirator.

THE MOUNTAIN LION IS FOUND ACROSS SUCH A BIG RANGE – FROM CANADA TO FLORIDA – THAT IT'S KNOWN BY MANY DIFFERENT NAMES, INCLUDING COUGAR AND PUMA.

MOUNTAIN LION

This is North America's version of a big cat, although technically it's classified as a small one. With a slender body up to 2.4m from head to tail and a weight up to 65kg, it's not nearly as big as the lions and tigers of Africa and Asia, but is powerfully built and an equally impressive hunter. Mountain lions are nocturnal predators that hunt by silently stalking prey, then pouncing on it and grabbing it with their muscular front legs. These powerful limbs are adapted to bring down prey as large as elks or moose, and their strong hind legs help them jump almost 6m from the ground. They can't roar, instead they purr like a house cat.

BLACK BEAR

This is North America's most common bear species. Although black bears have attacked and killed people, they are not considered to be as much of a threat to people as the larger, more aggressive grizzlies are. They can be dangerous when they become used to finding their food near where humans congregate though.

EYES DOWN

If you do encounter a bear, avoid making eye contact, as they will regard this as aggressive and treat it as a challenge.

EASTERN CORAL SNAKE

The pattern of the eastern coral snake has come to be associated with such a toxic bite that other snakes have evolved to look like it as a form of protection. Their venom contains potentially fatal toxins, but there haven't been any human deaths from this snake since an antivenom was developed.

SLOW RELEASE

It can take more than 12 hours for a bite victim to start feeling the full impact of the eastern coral snake's venom.

COYOTE

People often hear the howl of coyotes, as they live around cities to scavenge for food scraps. Attacks are rare, and they're not usually seen, as coyotes are nocturnal and secretive. Away from humans, they'll hunt on their own or with a partner if after bigger prey. Occasionally, when chasing bigger animals such as deer, they'll hunt in a pack.

KEEP SWIMMING

Coyotes are very good swimmers – they've even managed to colonise islands off the coast of north-eastern USA.

GREAT WHITE SHARK

The world's biggest predatory fish, the great white shark is found in cooler waters on both sides of the North American continent (and in oceans around the world). However the threat they pose to people is surprisingly small – there's a far greater risk of being killed by a bee or a lightning strike. Each year, on average, there are about 19 shark attacks in North American waters, including one fatal attack every two years. Scientists believe most great white attacks are primarily due to the animals being curious. The trouble is that when you're as big as a great white, with such huge jaws and multiple rows of sharp teeth, any bite can cause major, possibly fatal, injuries.

DEADLY FACT!

Adult great whites usually grow to around 4.5m and weigh more than two tonnes, but records show they can reach a length of 6m!

JAGUAR

Jaguars kill their prey in a way that no other big cat does, going for the head and biting directly through the skull to pierce the brain. It's a highly effective killing method, but can leave jaguars with broken teeth. These cats are stealth hunters that stalk their prey before suddenly bursting out from cover to kill it, and their spotted coats make great camouflage in the dense rainforests of South and Central America, further enhancing their hunting skills.

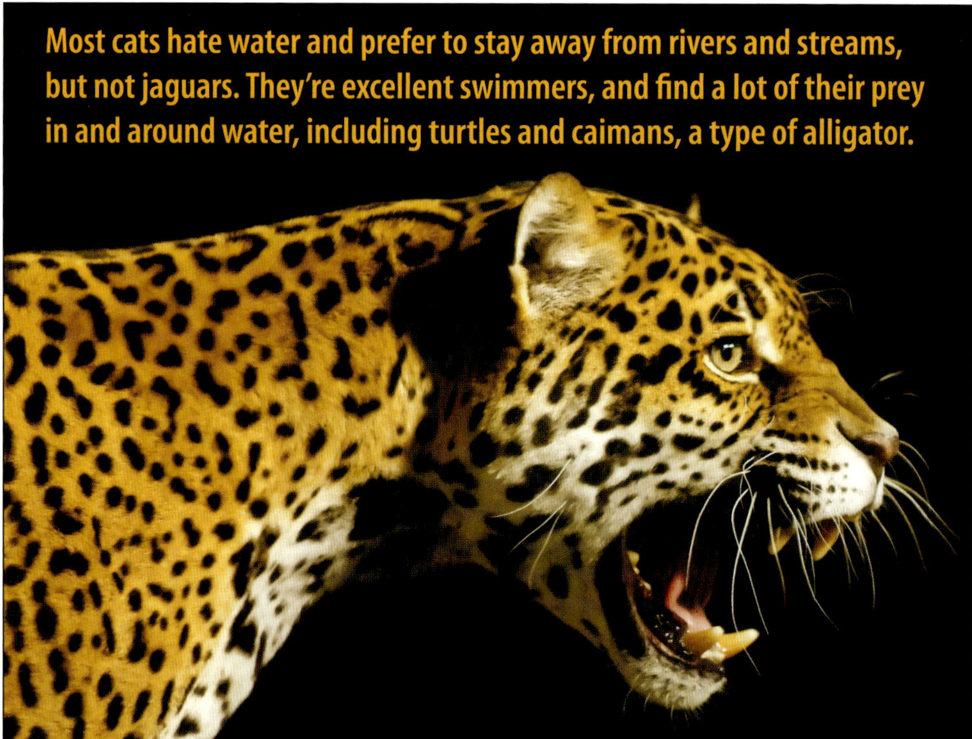

Most cats hate water and prefer to stay away from rivers and streams, but not jaguars. They're excellent swimmers, and find a lot of their prey in and around water, including turtles and caimans, a type of alligator.

POISON DART FROG

There are more than 100 different species of poison dart frog in Central and South America. Most are brightly coloured, and the stunning blue, red and yellow hues of their skin warn that they are poisonous and best avoided. Native South Americans have been using this to their advantage for hundreds of years, by dipping their hunting darts in toxic secretions from the glands on the backs of these frogs.

WEIRD BUT TRUE

When poison dart frogs are raised in captivity they aren't always poisonous, unless they are fed on the same creatures they would eat in the wild.

ELECTRIC EEL

This freshwater fish can produce large amounts of electricity, which it uses to stun prey, as protection from predators, and to communicate with other electric eels. They are able to produce electrical bursts of up to 600 volts, which is enough to stun or even kill a human. These fish produce electricity using three special organs. Inside these are thousands of specialised cells called electrocytes, which function like tiny batteries.

It's thought that many people have drowned in the Amazon after receiving an electric shock from this eel.

PIRANHA

Piranhas are freshwater fish famed for their razor-sharp teeth and aggressive attitude. They're known for working together in a frenzied group to bring down prey that is considerably bigger than them, which includes large mammals – and occasionally a person who made the mistake of swimming in piranha-infested waters. They school in groups to use the safety-in-numbers approach to protecting themselves from predators. Most are carnivores or omnivores, but a few are actually herbivores. There are up to 60 different species of piranha, including the red-bellied piranha, which has the strongest jaws and sharpest teeth of all. They are all found naturally only in the lakes and rivers of South America.

RARE OFFENDERS

Although people have died from piranha attacks in the Amazon River, fatalities are rare.

BRAZILIAN WANDERING SPIDER

These spiders are big, aggressive, and very deadly. There are eight different species of Brazilian wandering spider, and some have a leg span of more than 15cm! They're found in jungle habitats, where they hide during the day under rocks or logs, inside termite mounds or in plants. At night they come out to wander across the forest floor looking for prey. Their venom contains a nerve poison, and although the killing power of the species varies, they are all generally considered to be amongst the world's most venomous spiders.

The scientific name for this spider, phoneutria, means murderess.

ASSASSIN BUG

This large insect gets its name from the way it kills its prey. Often approaching from behind, it stabs the victim and injects an enzyme that dissolves its inner organs, then feeds on the fluid. It feasts on reptiles, insects and birds. While most bugs that feed this way have two separate tubes, one for injecting and one for sucking, the assassin's beak acts as both syringe and siphon. This allows it to target prey twice its own size.

BLACK CAIMAN

Not only is this reptile the largest member of the alligator family, the scaly beast is also the largest predator in the Amazon River basin, and its skin colouration is great camouflage when it hunts its prey at night. Unfortunately it's been known to attack humans, and has occasionally killed them.

WIPED OUT

These prehistoric animals were almost hunted to extinction due to high demand for their skin, but have rebounded.

SOUTH AMERICAN RATTLESNAKE

The bite of this snake is potentially deadly, and if it doesn't kill you, it can still cause permanent disabilities. It's found mostly in dry or arid habitats, and usually preys on mammals, although it will also catch and eat lizards. South American rattlesnake venom contains potent nerve poisons that can cause a victim to become progressively paralysed. This effect can be so strong that the neck becomes limp and appears to be broken, making breathing difficult, and then impossible, without medical intervention. Blindness, which can often be permanent, is another nasty symptom.

GOLDEN LANCEHEAD

This tree-climbing snake is only found on Queimada Grande, a tiny island off the coast of Brazil. It's a type of pit viper, a family that contains the dreaded rattlesnake, which all have special pits on their heads, between the eyes and the nostrils, that detect heat and infrared radiation. This means they can detect the warmth given off by the bodies of their prey at night. They can do this from up to a metre away, which is like being able to see in the dark.

This snake likes to eat birds and mammals, but will take lizards if they come too close.

TERCIOPELO

Another pit viper, this is the most dangerous snake of Central and South America, causing more human deaths than any other American reptile. Its pits allow it to identify a rise or drop in temperature of just 0.001°C, helping it detect warm-blooded mammals. Coupled with its tongue, which it uses to taste the air, it can hunt and strike even in total darkness.

The Jackson's chameleon – also known as the horned or three-horned chameleon – is native to East Africa, but now lives in the US too. The males put on incredible colour displays to woo females.

FREAKY

Bizarre body shapes, unusual hunting techniques and extraordinary talents and lifestyles are just some of the reasons that these animals rule!

AYE-AYE LEMUR

With its transfixed stare and wild fur, the critically endangered aye-aye has long been considered an omen of bad luck by the people of its native Madagascar. Related to chimpanzees, apes and humans, these metre-tall primates use their sharp claws and opposable big toes to dangle from branches in rainforest trees. While perched there, the aye-aye raps its long middle finger on the bark and listens for wriggling insect larvae in the wood. It fishes them out with the same pointed claw, which is also used for scooping the flesh out of coconuts. The aye-aye is the only primate known to use echolocation – calling out and listening for an echo to determine location – to find dinner.

IN GOOD CONDITIONS – SADLY LESS LIKELY IN MODERN TIMES – THE AYE-AYE CAN LIVE TO 23!

SATANIC LEAF-TAILED GECKO

With its demonic red eyes, angular horns and wicked smile, it's an easy guess as to how the satanic leaf-tailed gecko got its name. Native to the rainforests of Madagascar, it also has an alarming bright red mouth that it uses to warn off predators. If you don't scare easily, then consider the gecko's other defence mechanism – camouflage. It would be easy to think you were picking up a fallen leaf instead of the tail of this tiny gecko, but if you did, it would likely drop its tail and scamper away.

FREAKY FACT!

The satanic leaf-tailed gecko doesn't just look like a leaf, it acts like one too, spending most of the day hanging motionless off a branch or curled among leaf litter.

STRIPED LEAF-NOSED BAT

Blessed with a face only a mother could love, the striped leaf-nosed bat lives in colonies of tens of thousands inside caves in eastern and southern Africa, and occasionally roosts in trees and under the eaves of buildings. At dusk they hunt, using echolocation – calling out into the sky and listening to the echoes that come back to them – to find their beetle prey. It's believed that their unusually shaped nose helps with this process, but how that actually works remains a mystery.

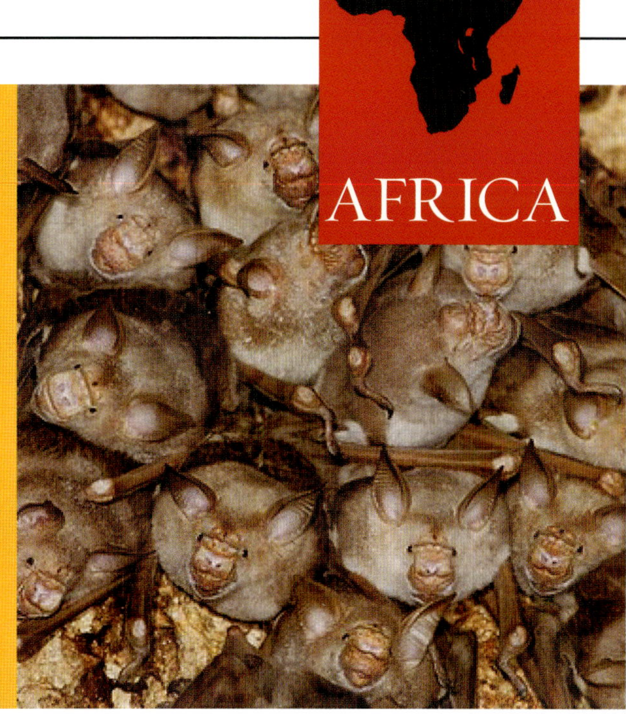

SOCIABLE WEAVER BIRD

Common in the Kalahari Desert region of southern Africa, these finch-sized birds build the largest tree nests in the world, measuring up to 6m wide by 3m tall. Their nest is like a massive unit block housing up to 100 families – and sometimes, nests become so heavy they topple the tree! From a distance you could mistake them for haystacks hanging in a tree, and some nests are used for more than 100 years.

THIRSTY MUCH?

Living in the desert, these birds have adapted to never needing to drink water – they get all the moisture they require from eating bugs.

HONEY BADGER

Strong, smart and fierce, honey badgers, or ratels, have coarse black-and-white coats that might remind you of a skunk. It's not just their appearance that's similar either. When threatened, honey badgers release a stink bomb that repels predators such as lions, leopards and hyenas. If that doesn't work, their skin is really thick, which means it's hard to pierce, and also loose, so they can twist and turn within it, allowing them to bite their attacker.

Honey badgers have a strong immunity to the venom of snakes and scorpions – which is very useful, because they are their favourite meal!

ARMADILLO GIRDLED LIZARD

Measuring just 21cm in length and covered in spiny scales, the armadillo girdled lizard rolls into a ball, tail in mouth, underbelly protected, when threatened. Native to the western coast of South Africa, this fierce-looking lizard has such a powerful bite that it's been known to break its own jaw while eating. It lives in large family groups and eats bugs and spiders.

FREAKY FACT!

These lizards are really sociable, and unlike most others, they will live in groups instead of on their own – even when there is plenty of separate shelter to choose from!

GERENUK

This long-necked antelope has muscular hind legs that allow it to stand upright. By using its front legs to bend branches downwards, it can nibble leaves that grow up to 2.5m off the ground. It also has rabbit-like ears, which are always listening for predators. These antelopes use sounds to communicate – a buzzing when alarmed, a bleat when in extreme danger, and a whistle when annoyed.

OH DEER!

A gerenuk can go its whole life without water, getting all the moisture it needs from its diet of leaves, flowers and fruit.

GIRAFFE

No other land animal on earth is as tall as the giraffe. If you were on the second floor of a building, a giraffe could easily gaze in on you through the window. Its neck is as long as its legs, about 1.8m, and its feet are the size of a dinner plate. In fact, everything about the giraffe is oversized – its heart is 60cm long, and its black tongue stretches up to 50cm long and can grasp things.

COAT OF ARMS

The pattern of a giraffe's coat is unique, just like a human fingerprint. The shape of the patterning reflects where it lives in Africa.

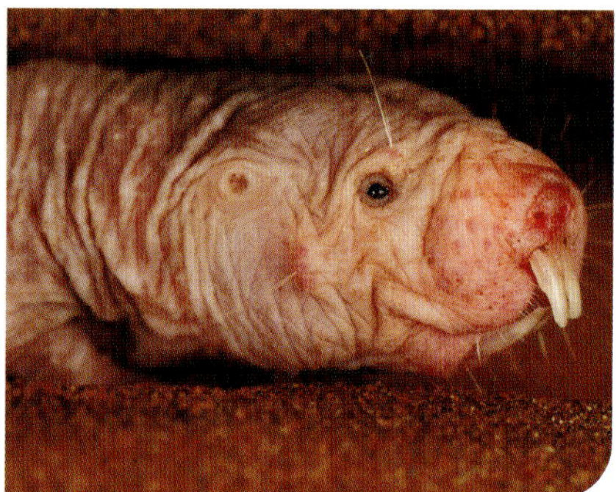

NAKED MOLE-RAT

Looking like a cross between an uncooked sausage and a miniature walrus, this animal can fit in a teacup. Also known as sand puppies, these virtually hairless rodents live in colonies of around 75, in an area around the same size as six football fields, and rarely leave the safety of their sandy burrows. Like bees, the colony is led by a queen, and the foot soldiers dig, collect food and tend the young.

YOU EAT WHAT?

Naked mole-rats eat roots and bulbs, but have also been known to eat their own poo to collect extra nutrients, and can even chew through concrete.

EUROPE

OLM

There's nothing normal about this slippery salamander – it's eyeless so is blind, is only found in underground rivers, swims like an eel, can go 10 years without eating, and lives to 100. And that's just to start with! The olm spends its whole life underwater, and finds food using sensors in its snout. Females lay eggs every five to six years, up to 70 at a time, on the ceiling of a cave. And she can smell if the young inside are growing or not – and if not, she eats them!

FREAKY FACT!

Native to Central and Southeast Europe, the olm is listed as vulnerable. Your best chance of seeing it is in Slovenia's Postojna Cave, where underwater cameras and TV monitors show its secretive aquatic antics. Centuries ago, when floods washed them from their caves, people believed they were baby dragons!

WALRUS

Tusks, moustache-like whiskers, wrinkled pinky-brown skin and a blubber-filled body combine to make this 3.5m-long marine mammal unique. It uses its tusks to pull itself out of the water, earning it the name tooth-walker, and also to break breathing holes in the Arctic ice from underneath. The tusks can grow up to a metre long, and males also use them in fights over territory and females.

HANDY HAIRS

A walrus's whiskers are called vibrissae, and they help it feel for food, especially shellfish, on the dark ocean floor.

RUFF BIRD

It's the puffed-up neck feathers that give this wading bird its name – a ruff was a fancy ornamental collar worn by Europeans in the 16th and 17th centuries – yet only breeding males display it. They also develop a tuft of brightly coloured feathers on their head, and orange legs, beak and facial skin. But as soon as mating has finished, the fancy feathers moult and the face, beak and legs become duller in colour, and the males leave the females to incubate the eggs then raise the chicks alone.

FREAKY FACT!

Known for aggressive displays at communal mating grounds, ruffs used to go by the scientific name *Philomachus pugnus*, which means battle loving, but it has since been updated.

EUROPEAN RIVER LAMPREY

Like something from a vampire movie, this jawless species attaches itself to larger fish to drink their blood. Its mouth contains two rows of circular teeth to help it cling to the host as it swims. It has one nostril and seven breathing holes along its body.

SPAWN AND DIE

Similar to salmon, these fish migrate from coastal waters to inland breeding grounds to spawn. They swim upstream in autumn and winter, lay eggs in the spring, then die. When the young mature, they head to the sea.

EUROPEAN BADGER

Like bears, European badgers hibernate in the winter, settling into setts, a network of tunnels and chambers often passed down through generations, to wait out the big chill. During the summer, they put on large amounts of body fat to nourish them through the cold dark months underground. Up to 12 animals might cram into each sett, which has a series of tunnels running off it to above-ground openings, and may extend for 100m or more. This kindly creature is a lover not a fighter, and will share its sett with other animals, including rabbits, otters, red foxes and pine martens.

European folklore and fairytales are filled with badgers, although they're not always portrayed as friendly. In *The Wind in the Willows*, Badger is a wise but grumpy fellow who "simply hates society," while an evil badger called Tommy Brock kidnaps Benjamin Bunny's children in Beatrix Potter's *The Tale of Mr Tod*.

SHORT-SNOUTED SEAHORSE

With the head of a horse, a long forward-curling tail, and a set of spiny eyelashes, the seahorse is an unlikely looking creature. Short-snouted seahorses measure just 13cm from head to tail, and are able to change colour – from green and yellow to maroon, purple and black – to mimic the plants in the shallow waters where they live.

NASTY BEHAVIOUR

Males are aggressive when fighting for a mate – they use their snouts to shove their rivals, and their tails to wrestle one another. But once partnered, they mate for life.

DIVING BELL SPIDER

This clever little arachnid only needs to come to the surface of its pond once a day to breathe – it spins a web underwater and fills it with oxygen. Using the fine hairs on its abdomen, it moves tiny bubbles of air from above the water surface to its underwater diving bell. The bubble works a bit like a fish's gills, taking oxygen from the water and sending carbon dioxide back out.

DIVING BELL SPIDERS ARE THE ONLY SPIDER SPECIES THAT LIVES THEIR ENTIRE LIVES UNDERWATER.

OCEAN SUNFISH

One of the ocean's true oddities, this is the world's largest bony fish, measuring 1.8m long, and 2.4m between top and bottom fin tips, and weighing up to 1000kg (although one weighing 2250kg has been spotted!). Its bullet-like shape is the result of a tail fin that never grows, instead folding in on itself and forming a rudder called a clavus. Often seen sunbaking near the ocean's surface, its huge dorsal fin cuts through the water, and is often mistaken for a shark's. But don't fear – this giant fish only has a taste for jellyfish, zooplankton and algae.

PESKY PARASITES

Sunfish can be overrun with parasites, so they jump up to 3m out of the ocean, landing with a slap to try to get rid of them.

GREATER FLAMINGO

This regal bird owes its pink colour to the animals it eats, shrimp-like crustaceans that live in the saltwater mudflats of southern Europe. Stirring up the mud with its webbed feet, the flamingo buries its long bent beak – and sometimes its whole head – in the water to suck up the tiny treats. Its tongue pumps up and down, pushing the water out of its mouth and trapping the food in tiny filters. At 1.5m tall, the greater flamingo is the largest of the flamingo family, and the most widespread.

FREAKY FACT!

Greater flamingos live and breed in colonies of up to 200,000. There's safety in numbers – while some birds stand watch, others can feed. A loud, deep warning honk, similar to that of a goose, alerts them to predators.

ASIA

PROBOSCIS MONKEY

No other primate has such a large, fleshy, dangling nose as the male proboscis monkey. It's believed that their nose helps increase the volume of their call, impressing females and intimidating rivals. Proboscis monkeys have evolved webbed feet and hands to swim quickly, a great help when evading the crocodiles that lurk in the rivers of their native Borneo. They can't eat ripe fruit, as the sugar ferments and expands in their bellies, causing them to bloat. This can be deadly, so they eat unripe fruit as well as leaves and seeds. Habitat destruction throughout Borneo means this big-nosed primate is endangered.

QUICK ESCAPE

With a body the size of a rabbit, and legs as thin as pencils, the lesser mouse-deer is just the right size to scamper through the undergrowth of South-East Asia's tropical forests.

LESSER MOUSE-DEER

The smallest known hoofed mammal, the lesser mouse-deer grows to just 45cm long and weighs a mere 2kg. At first glance, it appears a mish-mash of deer and mouse, but it's actually neither. It's a member of the tragulidae family, which means tiny goat. Generally docile, and threatened by predation by feral dogs, males will angrily beat their hooves when alarmed, stomping on the ground up to seven times per second. This drum roll usually scares away predators, but if forced to fight, they will use their tusk-like canine teeth to tear at their foe.

TUFTED DEER

By far the scariest looking deer, this vampiric Bambi has oversized protruding canine teeth that look like fangs. Males use these 2.5cm-long tusks during mating season to fend off rivals. When threatened, these animals bark loudly then flee in cat-like leaps. Named for the distinctive patch of coarse hair on their forehead, which hides their small antlers, this deer lives in high-altitude forest regions in north-east Myanmar (formerly known as Burma) and southern and central China.

A timid creature, if disturbed, the tufted deer will flash the white underside of its tail, which confuses predators just long enough for it to escape.

EMEI MOUSTACHE TOAD

Mating season for the Emei moustache toad is a macho affair. In February and March each year, males sprout up to 16 spines on their top lip, and migrate from the forest to swift-flowing rivers to breed. They compete for territory and females by head-butting each other's bellies, often causing puncture wounds. When mating is done and the eggs have spawned, the spines fall out.

In the frog world, male frogs that grow larger than their female counterparts (as is the case with the Emei moustache toad) are known to be aggressive. Most of the time in frog species – 90 per cent, in fact – females are larger.

GREAT HAMMERHEAD SHARK

It's obvious how this shark got its name! Its wide-set eyes and mallet-shaped head are used to find and attack fish, nailing them to the sea floor. Covering its expansive snout are thousands of electroreceptors called the ampullae of Lorenzini, used to detect the electrical fields created by prey animals. This is useful for hunting down its favourite meal, stingrays, which bury themselves in the sand. There are nine species of hammerhead shark, but only the great hammerhead is considered dangerous.

DID YOU KNOW?

Swimming above the ocean floor, hammerheads swing their T-shaped heads from side to side, like you would a metal detector on a beach, looking for prey.

COCONUT OCTOPUS

The only invertebrate known to use tools, and one of only two octopuses known to "walk" on two of its legs, the coconut octopus is named for its habit of carrying coconut shells across the sea floor and using them to build fortresses. Before construction can begin, the octopus must dig up and thoroughly clean its building materials – during this time the shells afford no protection, leaving the cephalopod open to attack.

WREATHED HORNBILL

This distinctive bird mates for life. It has a breeding practice akin to an ancient Egyptian burial – when the female is ready to lay her eggs, she builds a nest in a tree cavity. The male then entombs her in the cavity, sealing it with mud, fruit and faeces and leaving only a slender hole through which he feeds her and their chicks. For around four months, the female wreathed hornbill and her offspring depend solely on the male for their survival.

MIMIC OCTOPUS

Using colour-matching and shapeshifting to avoid predators, this clever octopus imitates a range of creatures – including a lionfish swimming with its spines erect. Perhaps its most incredible morph is when imitating the venomous banded sea snake. It conceals six of its arms in the sand then raises the other two, now coloured with thick black and beige stripes, in opposite directions in order to resemble the reptile.

FLYING DRAGON LIZARD

Soaring on outstretched wings, flying dragon lizards move effortlessly around their native forests in South-East Asia and India. The wings are actually skin extensions called patagia, which allow the lizards to easily glide through the air. The maximum flight trajectory for both males and females is 8m, which is 40 times their body length. Their patagia also helps identify different species, as each of the 45 known species of flying dragon lizards has a unique colour display.

WHILE SOME LIZARDS ALSO HAVE SMALL FLAPS OF SKIN ALONG THEIR RIBS, MEMBERS OF THE FLYING DRAGON GENUS ARE THE ONLY KNOWN REPTILES RECOGNISED AS TRUE GLIDERS.

PAEDOPHRYNE AMAUENSIS

At just 7mm long and the size of a fly, these microhylid frogs are the world's smallest vertebrates. Found in the eastern rainforests of Papua New Guinea, they live in leaf litter on the forest floor, and eat tiny invertebrates such as mites, which are ignored by bigger predators. They're adept jumpers, and can leap 30 times the length of their body. Their high-pitched call sounds like an insect, and is difficult for humans to hear – which is perhaps why they remained undiscovered until as recently as 2009. They're also too new to have a common name yet.

Unlike other frogs, this one's life cycle doesn't have a tadpole phase. Instead, they hatch as "hoppers", miniature versions of the adults.

PARADISE FLYING SNAKE

In the jungles of South-East Asia, these fascinating reptiles, also known as paradise tree snakes, use free fall and contortion to "fly". Starting by resting on the end of a branch, they flatten their body to about twice its normal width, forming a parachute shape that traps air, then propel themselves forward. To turn, they wriggle back and forth. These snakes, which can have green, yellow or orange spots, are one of the smaller flying snakes, at only 60cm in length, and the best gliders. They've been recorded travelling up to 100m through the air.

OCEANIA

LEAFY SEA DRAGON

Dripping with "leaves", these sea dragons blend perfectly with the seaweed and kelp forests off southern Australia where they live. Closely related to seahorses, they drift with the currents in search of microscopic prey, like sea lice. Graced with thin, tubular snouts and small, leaf-shaped fins, they are usually browny-yellow in colour, and 35cm from top to tail.

FREAKY FACT!

A bright yellow tail on a male leafy sea dragon is a sign he's ready to mate. Like seahorses, males are in charge of raising their offspring. Unlike seahorses, which carry their young in a pouch on their stomach, male sea dragons nurture theirs on the underside of their tail.

MARY RIVER TURTLE

With a 50cm-long carapace, the endangered Mary River turtle is one of Australia's largest. But it's not its size that's most interesting, it's the fact that it uses its enormously long tail to breathe. This unusual feature means the turtle breathes through its back passage, and has another less-polite name, the "bum-breathing turtle". It also has long whisker-like barbels on its chin, which hold its taste buds and helps it feel out food.

FREAKY FACT!

In turtles, as for other reptiles, the back passage is called a cloaca. Urine and faeces, as well as eggs, pass out through there.

TREE KANGAROO

Unlike their ground-dwelling cousins, tree kangaroos have long muscular arms and short legs. They also have curved nails and spongy pads that help with gripping tree branches, while their long tail helps with balance. Graceful climbers, they wrap their arms around tree trunks and hop upwards using their hind legs. They are expert jumpers too!

NO SWEAT

The 14 species live in the rainforests of New Guinea or north-eastern Australia, and they keep cool by licking their arms.

GIANT PRICKLY STICK INSECT

Covered in thorns, the female giant prickly stick insect is intimidating. At 20cm long, she's twice as big as a male, and sprays an odour to scare off predators. While males do this too, they're more likely to fly away from danger, but the female's wings are sadly too small, and her body too heavy, for flight.

HOME AMONG THE GUM TREES

These huge insects live in eucalypt forests in north-eastern Australia, and can be found in a variety of earthy colours.

SUPERB BIRD OF PARADISE

When an adult male superb bird of paradise wants to attract a female, he fans out the velvety black feathers on his back, and the gleaming blue feathers on his chest to form a spectacular cape. He then begins snapping his tail feathers and hopping around in a fancy dance, hoping a female will partner with him.

HEAVENLY CREATURES

These stunning birds live in the forests of New Guinea. The local name for them, bolon diuata, means birds of the gods.

GUINEAFOWL PUFFERFISH

With its torpedo-shaped body, big head and large eyes, this fish is best known for its talent for inflating like a balloon to scare off predators. To do this, it drinks lots of water to fill its extremely elastic stomach. But this isn't its only defence. It's also covered in spines, and is highly poisonous if eaten. The guineafowl pufferfish can be either bright yellow, or black with yellow spots – the dark ones are more toxic.

The pufferfish has four large constantly growing teeth that are joined together to form a beak-like structure. It uses them to grind down the squid, krill, clams and crabs that it enjoys eating.

CASSOWARY

Standing as tall as a person, the world's most dangerous bird has a scaly blue head, wattles (red flaps of skin that hang from its throat), and atop its head an eye-catching casque, or helmet, made of toughened skin, which is hard on the outside and spongy inside. But most threatening is the 12cm-long inner claw that is like a dagger, and can be used to disembowel some predators. Males raise the young, and will protect them at any cost.

FREAKY FACT!

A shy creature, the cassowary is an elusive quarry. People can search for years and never spot one, or even catch a glimpse of it as it hightails it into the rainforests of north-eastern Australia, at speeds of up to 50km/h.

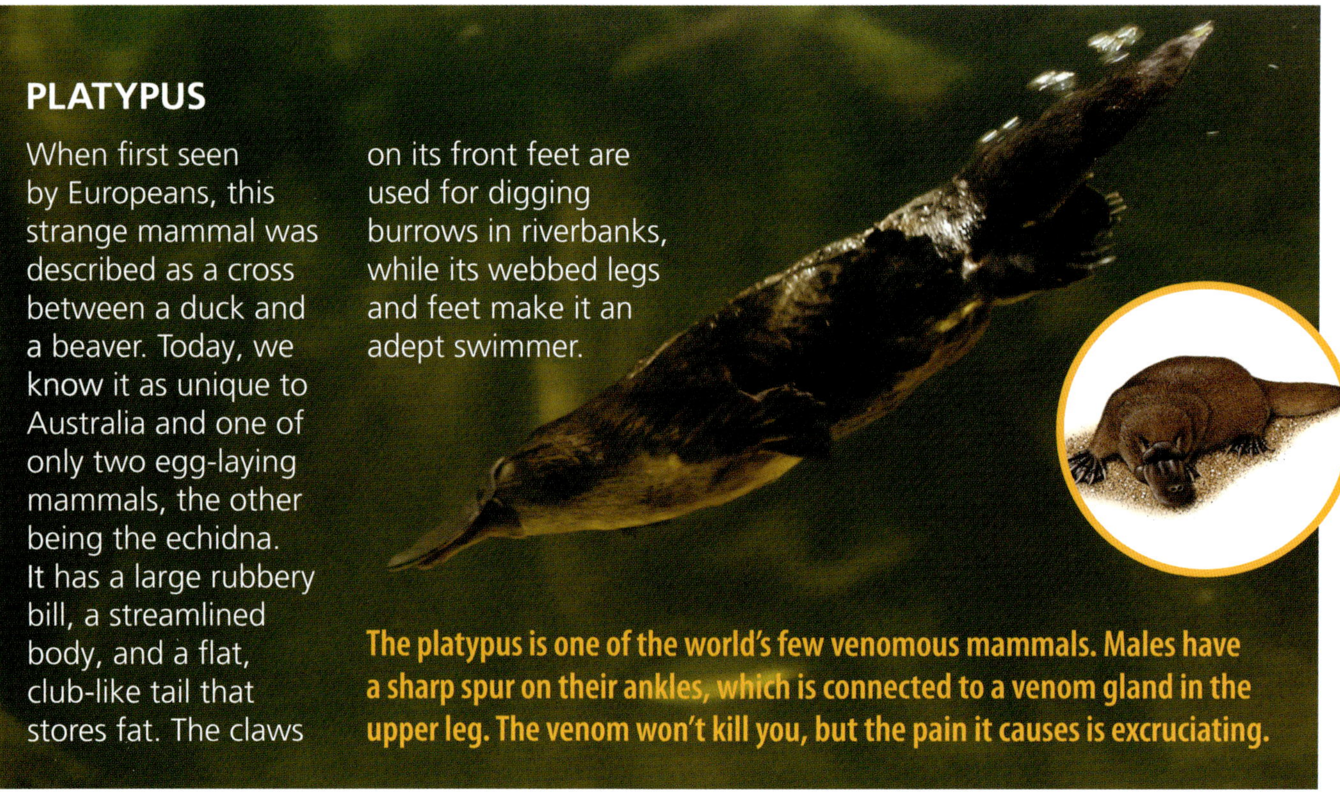

PLATYPUS

When first seen by Europeans, this strange mammal was described as a cross between a duck and a beaver. Today, we know it as unique to Australia and one of only two egg-laying mammals, the other being the echidna. It has a large rubbery bill, a streamlined body, and a flat, club-like tail that stores fat. The claws on its front feet are used for digging burrows in riverbanks, while its webbed legs and feet make it an adept swimmer.

The platypus is one of the world's few venomous mammals. Males have a sharp spur on their ankles, which is connected to a venom gland in the upper leg. The venom won't kill you, but the pain it causes is excruciating.

WRAP-AROUND SPIDER

Coloured to conceal and happily overlooked, this spider is perfectly built for tree hugging. Its body is shaped like a shield, with the underside concave to fit snugly around a branch, and its abdomen slightly raised to complete the line. Some even have bumps or turrets on their back so they look just like a tiny twig.

QUICK ESCAPE

Disguised by day, this Australian native builds large orb-shaped webs between trees at night to catch its insect prey.

CUTTLEFISH

This oceanic master of disguise changes colour depending on its surroundings. It has eight arms, two tentacles, three hearts, and one of the largest brains compared to its body size of all the invertebrates. It's not a fish but a mollusc, and it has a sharp, beak-like mouth that it uses like scissors to cut open flesh, before using its tentacles to tear out the meat.

You've probably found pieces of white feather-shaped cuttlebone washed up on the beach. It's unique to cuttlefish, and is what keeps it afloat. Jewellers use it to make moulds, and it's often given to pet birds as a source of calcium.

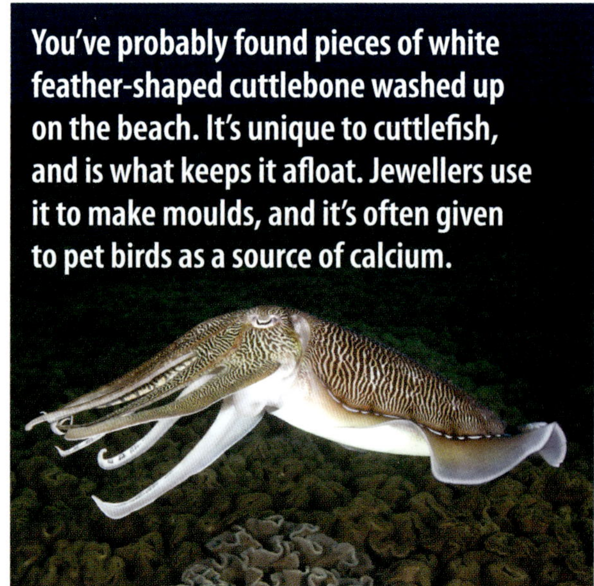

GLASSWING BUTTERFLY

NORTH AMERICA

While most butterflies have colourful, patterned wings, designed to warn off predators, this species has chosen invisibility as its defensive cloak – its wings are mostly transparent. But that's not all that's amazing – this butterfly's 6cm-wide wings can carry roughly 40 times its own weight. It's also exceptionally fast, flying at speeds of up to 13km/h.

DID YOU KNOW?

The glasswing butterfly's Spanish name is espejitos, which means little mirrors.

BURROWING OWL

These ground-dwelling owls nest in burrows dug by other animals, including squirrels, armadillos, skunks and gophers, in the open plains throughout North America. Galloping across the ground on their long legs, they pounce on prey, including insects, small mammals, amphibians, reptiles and other birds, around dawn and dusk. Left on their own in the burrow while the parents hunt, juvenile owlets scare off predators by mimicking the sounds of a rattlesnake.

BEETLE JUICE

Burrowing owls surround their nests with mammal waste. The heady concoction attracts dung beetles, one of their favourite foods.

SOLENODON

The elongated snouts of these small burrowing, insectivorous mammals are unusually flexible, which means they can twist and turn their snouts to probe tiny nooks and crannies for food. The two living species are the Cuban solenodon and the Hispaniolan solenodon, which are endemic to their respective islands. Rare amongst mammals, their saliva is venomous, which helps them catch prey. Both species are sadly endangered.

NORTH AMERICAN PORCUPINE

This prickly fellow wears a coat of quills – a pointed warning that it's not easy prey. Its needle-like spines lie flat until it's threatened, and then they bristle on command. This porcupine is the largest member of its species, reaching more than 1m in length, and has more than 30,000 barbed quills. Each one has a sharp tip, and is difficult to remove once stuck in a predator's skin. For every quill that's lost, another one grows to take its place.

AXOLOTL

Also known as Mexican walking fish, axolotls are not fish at all, but amphibians. Like frogs and toads, they breathe through their skin and their gills – they have three on each side of the head, just above the legs. But unlike other amphibians, they don't develop past the larval stage – the tadpole phase in frogs – instead of taking to the land, the adults remain aquatic and gilled. Native to Mexico, axolotls are critically endangered.

An axolotl can heal almost any injured part of its body, regenerating arms, legs, tail, skin, and even major organs such as the heart, liver and kidney.

NORTH AMERICAN BEAVER

The first thing you'll notice about these beavers is their very distinct teeth. Large and strong, they're used to gnaw down trees and build dome-shaped homes, called lodges. And their powerful webbed back feet and paddle-shaped tails help them swim at up to 8km/h. They can also stay underwater for up to 15 minutes at a time, and have a set of transparent eyelids that function much like goggles – cool, right? This is particularly helpful because the front door to a beaver's lodge is located underwater.

FLANNEL MOTH CATERPILLAR

A furry coat may have inspired its alternate name of puss caterpillar, but touching this creature is definitely a no-no. The hairy comb-over conceals small, exceptionally poisonous spines that stick in the skin and cause excruciating pain that lingers for up to 12 hours. If you do happen to pet a puss caterpillar, the best way to remove the spines is with sticky tape. Lay it over the site, then rip it off. This caterpillar morphs into a flannel moth, which also has a pretty hairdo, but is harmless in contrast.

If you do happen to get stung, symptoms include nausea, headache and dizziness, and the inflammation will last a few days.

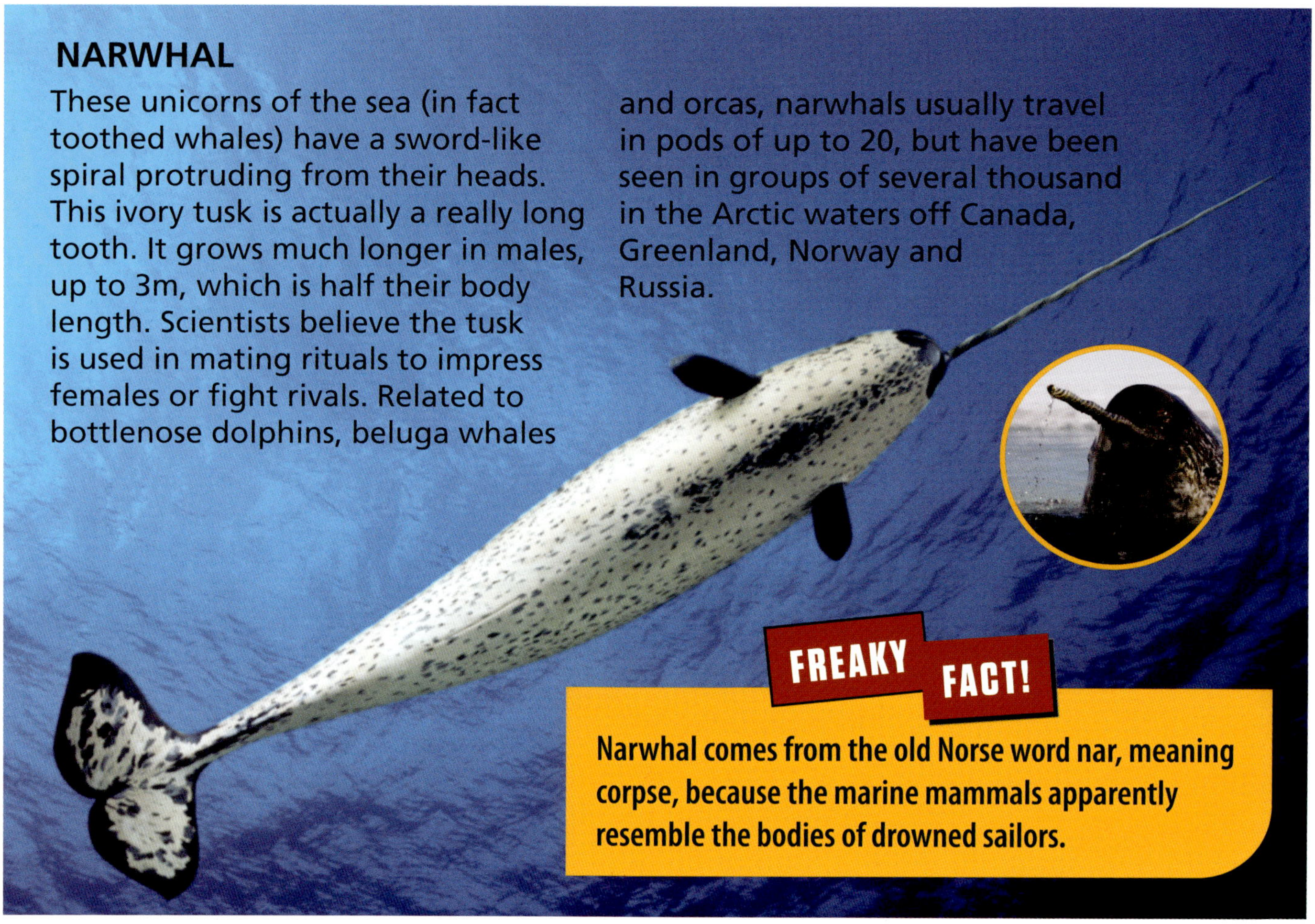

NARWHAL

These unicorns of the sea (in fact toothed whales) have a sword-like spiral protruding from their heads. This ivory tusk is actually a really long tooth. It grows much longer in males, up to 3m, which is half their body length. Scientists believe the tusk is used in mating rituals to impress females or fight rivals. Related to bottlenose dolphins, beluga whales and orcas, narwhals usually travel in pods of up to 20, but have been seen in groups of several thousand in the Arctic waters off Canada, Greenland, Norway and Russia.

FREAKY FACT!

Narwhal comes from the old Norse word nar, meaning corpse, because the marine mammals apparently resemble the bodies of drowned sailors.

STAR-NOSED MOLE

Resembling a strange blend of rat and octopus, star-nosed moles use their unique snout to find food. The 22 tentacles, which together have a diameter of 1cm, ring the nostrils and work as a tactile eye, feeling for prey. They're so quick that they've earned the award for fastest-eating mammal. This nose also lets them smell underwater. The moles blow bubbles, which mingle with the scent of a small fish or earthworm, then suck them back in – and then it's meal time!

TEXAS HORNED LIZARD

Wearing a crown of thorns, the tiny Texas horned lizard also goes by the name horny toad, because of its rounded body, which it puffs up when threatened. Its other defence mechanism is to shoot a stream of blood from the corners of its eyes. The liquid, which can travel up to 1.5m, is a cocktail of blood and chemicals that's foul-tasting to its predators.

SOUTH AMERICA

POTOO

Sitting motionless on the stump of a branch matching its colour, beak pointed skyward, these birds are easy to miss during daylight hours. At night though, their huge staring eyes and eerie call, "po-*too*, po-*too*", light up their South American forest homes. Swooping from their perches, they hunt for flying insects, swallowing them whole with their large, wide mouths.

LUCKY SONG

Of the seven species, the largest is the great potoo. It has a wingspan of 1m, and makes a unique moaning growl that's unsettlingly eerie when heard in the jungle at night. According to Brazilian legend, it's a song from the dead – giving good luck to friends and bad luck to enemies.

KINKAJOU

Also known as honey bears, because of their golden coat and appetite for sweet treats, kinkajous are arboreal mammals that use their long sticky tongues to raid bee hives and termite nests. They live in the tropical rainforests of Central and South America, and have rotating ankles that let them run up and down tree trunks without having to turn their body. Their strong prehensile (gripping) tail is as long as their 50cm body, and acts as a fifth limb, allowing them to hang upside down from branches.

SPREADING THE LOVE

Sometimes confused with monkeys, kinkajous help pollinate flowers. Pollen sticks to their faces and then smears off at the next blossom they visit.

PANDA ANT

Don't be fooled by the cute appearance of this insect – it's actually a wingless wasp whose sting packs a punch. Found only in Chile, the panda ant is boldly coloured to warn off predators. While males and females wear similar coats, they are often mistaken for different species, since the males are nocturnal but the females are not.

FREAKY FACT!

Female panda ants lay their eggs in the nests of ground-living insects, such as bees and wasps. When they hatch, the babies eat the developing larvae of the original insect.

YETI CRAB

White, hairy, and rarely seen, yeti crabs flock to thermal vents in the floor of the icy Southern Ocean to try to keep warm. Thousands of the crustaceans crawl over each other to get close to the boiling water without scalding themselves. There are three known species of yeti crab. The latest discovery, the Antarctic yeti crab, was found in 2010, and it warms itself up in waters of up to 400°C.

Because there's no sunlight where they live, 2.6km below the ocean's surface, they farm their own food. The tiny hairs on their shells grow bacteria, which the crabs then eat.

BASILISK LIZARD

Nicknamed the Jesus Christ lizard, this amazing animal runs on water. When threatened, it escapes by sprinting to the water's edge, then keeps going. The reptile runs on its back legs, body upright and arms by its side. It's so good on water because its big feet have flaps of scaly skin along the toes that create air bubbles, which it then pushes down on to keep afloat. As long as it moves quickly it won't sink. And its fast – it's been clocked at more than 8km/h on water and even swifter on land, speeding along at 11km/h.

BROWN-THROATED SLOTH

There's a lot that is unusual about this incredible mammal. Its thick brown fur often looks greeny-blue because of the algae growing on it, which helps it stay well hidden in its rainforest home. Instead of toes, it has three long claws on each foot, which help it hang upside down from branches. Its mouth is in a permanent smile, and most uniquely, it can swivel its head almost 300 degrees, thanks to its unique neck structure. The size of a cat, this sloth weighs around 4kg, and rarely walks – it stays up in the trees, except when it climbs down to poo once a week.

PINK FAIRY ARMADILLO

While this little fairy doesn't have wings, it is definitely enchanting. At only 15cm long, it's the smallest of all armadillos, has a fluffy white belly, and wears a coat of pink scales. The scales are so thin that they're almost translucent, and it's the blood vessels underneath that give them their rosy glow. Living on the sandy plains and arid grasslands of Argentina, it uses its claws to dig burrows to hide in.

SWORD-BILLED HUMMINGBIRD

The only species of bird to have a beak longer than its body, this sword-wielding hummingbird is perfectly designed to feed on flowers with tubular nectar chambers. Because its lance-like beak is so long, as is its tongue, it preens itself using its feet. From tail to beak tip measures 24cm, making it one of the world's largest hummingbirds. While it feeds, its wings beat really fast, rotating in a figure-of-eight movement rather than flapping, which holds it in place. Hummingbirds are the only birds that can fly backwards. This one is incredibly agile in the air, and performs beautiful flight displays, especially during mating season.

YURUANI GLASS FROG

There are no secrets with this little amphibian – the skin covering its belly and chest cavity is transparent so you can see its internal organs, such as the heart, and its bones. Found on the leaves of shrubs and trees growing along streams, this delicate 2cm-long frog also frequents waterways with brilliant red jasper rocks. Sticky pads and webbed toes allow it to lay its eggs on the underside of leaves that hang over the water. When hatched, the larvae fall into the stream below.

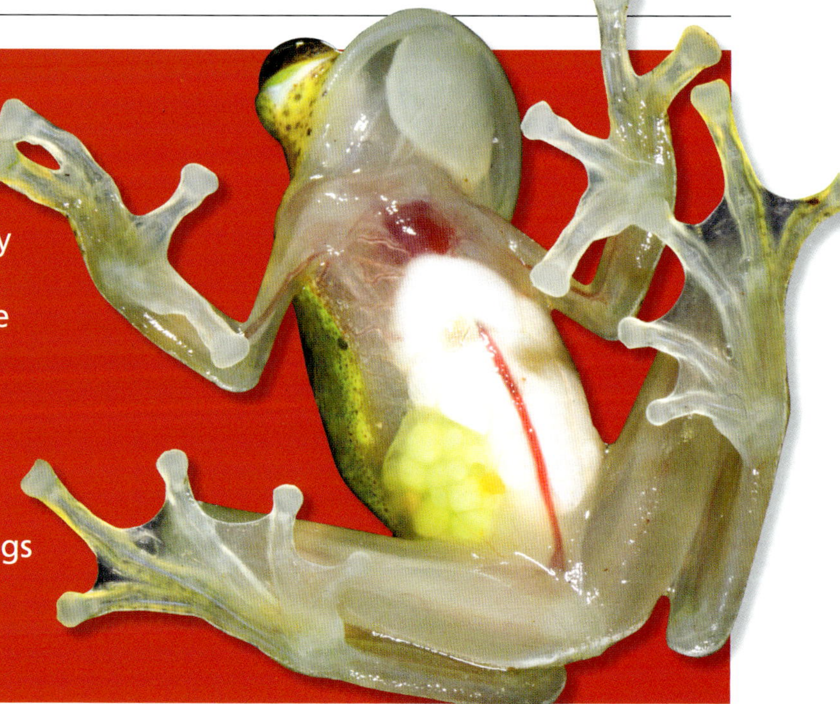

BLUE-FOOTED BOOBY

These showy seabirds are proud of their flashy feet – males goose step around the females during mating season. Clumsy on land, which is how they earned their Spanish name bobo, meaning stupid, these 90cm-tall birds are graceful in the sky. They're also incredible hunters. They wrap their large wings around their bodies and plummet into the sea from 24m above to pull out a fish. Roughly half of all breeding pairs of blue-footed boobies nest on the Galápagos Islands, off the north-west coast of South America.

BRAZILIAN PORCUPINE

With its oversized pinky-brown nose and soft, curious eyes, the Brazilian porcupine looks sweet enough to cuddle. But don't! It's covered in short, thick white-yellow spines, and has four long claws on each foot that can shred skin. The claws are designed to help with climbing – this animal spends more than three-quarters of its time in trees, eating, sleeping and socialising – and it has a long tail that it uses to curl around branches and grip on with.

A SPIKY CHARACTER

Like a toddler, this little mammal stomps its feet and cries when scared or upset. If confronted by a predator, it pops its quills, making it look twice as big. And if that fails, it rolls up into a spiky ball.

SURINAM TOAD

Looking like a leaf or a flat rock, this toad has a unique way of reproducing. The male pushes the fertilised eggs onto the female's back, and her skin grows over them. Once hatched, the tadpoles stay under there for up to four months – then when they're fully developed they push against the skin to loosen it, and burst free. The mother then sheds this skin, ready for the next breeding season.

FACT!

Also known as the star-fingered toad, this 20cm-long amphibian has a tiny star on the tip of each of the four fingers on its front legs. It lies with arms outstretched on the riverbed and uses them to feel for food – if something brushes against the stars, it sucks it up.

The Kleinmann's tortoise, also called the Egyptian or Leith's tortoise, is critically endangered and close to extinction.

ENDANGERED

The world's most endangered animals face a high chance of extinction within the next 50 years – or less – unless they can be saved by desperate conservation measures!

LEMUR

These primates are facing an extreme threat of extinction. There are 107 different species, and 33 of them are now critically endangered, including this red ruffed one. Another 70 are threatened, making them the world's most endangered mammals. Lemurs are found only on the Indian Ocean island of Madagascar, located off the east coast of Africa. They are illegally hunted, and suffer from habitat destruction. Their tropical forest homes are being logged for timber and cleared for agriculture – around 90 per cent of Madagascar's vegetation has been altered or destroyed.

ENDANGERED FACT!

Turning lemurs into Madagascar's number one tourist attraction is helping efforts to save them.

TARZAN CHAMELEON

There are many cases of animals facing extinction before we've had a chance to learn much about them, and this is another. Discovered in 2009, it was identified as one of the world's most endangered animals just three years later. It is found only on Madagascar, in two very small patches of forest that are both in a degraded state. Fortunately there are efforts underway to protect what's left of this precious lizard and its habitat.

WESTERN GORILLA

Gorillas share 98.3 per cent of their genetic code with humans. Sadly, they are another species on the endangered list because of illegal hunting. All four gorilla species are killed for food – their flesh is known as bushmeat – and some of their parts are used in medicines. Young gorillas are also captured for the pet trade. Their habitat in Africa is being destroyed too – their forests logged for timber and cleared for agriculture or new settlements – and another major risk is disease.

Since the early 1990s, ebola has caused waves of death in gorilla and chimpanzee populations in Africa. The virus kills more than 95 per cent of infected gorillas.

AFRICA

BLACK RHINO

The single biggest threat to the black rhinoceros is poaching. This illegal hunting is driven by a demand in Asia for rhino horns, which are used in Chinese medicine for a variety of ailments. However, their horns are made of keratin, the same protein as our own fingernails and hair, and there is no proof that they work against any disease or disorder. Sadly, rhino horns are also increasingly valued as symbols of wealth and success, pushing them further towards extinction.

DUE TO POACHING, BLACK RHINO NUMBERS CRASHED BY 98 PER CENT FROM 1960 TO 1995, TO LESS THAN 2500. THEY HAVE DOUBLED TODAY, THANKS TO CONSERVATION EFFORTS THAT WILL HOPEFULLY CONTINUE, BUT THEY ARE STILL AT RISK.

GOLDEN MANTELLA

Mantellas are tiny, bright frogs, just 2.4cm long. They're another animal group that's found only on the island of Madagascar. At least three mantella species are at risk of extinction, and each occurs in a very limited area and in habitat that is mostly degraded. Mantellas have skin toxins that they absorb from the ants, termites and other insects they prey upon, a form of defence that makes them distasteful and often poisonous to predators.

AFRICAN PENGUIN

The most distinctive feature of the African penguin is the patch of bare skin above each eye, which helps them cope with high temperatures in South Africa. The hotter the penguin gets, the more blood flows to these areas, where it's cooled by the surrounding air, decreasing the bird's temperature. Habitat loss, overfishing and coastal development continue to take a toll on this penguin population, their numbers declining by 60 per cent over the last 28 years.

ENDANGERED FACT!

The dot-like markings across each penguin's white chest are totally unique, with no other penguin having the same feather pattern.

PYGMY HIPPOPOTAMUS

Found in only four countries in West Africa, there are less than 3000 recorded pygmy hippos remaining. The forests that shelter them are being cut down or burned away, and the rivers where they swim are now polluted by humans. In logged areas, humans hunt them for their meat. These rare hippos have adaptations for spending time in the water, but they are much less aquatic than the common hippo, leaving them even more vulnerable to being preyed upon.

SPIDER TORTOISE

The spider tortoise is easily recognisable by its distinctive yellow shell with a black web-like pattern. It is the smallest of the four tortoises found in Madagascar, and is believed to live for up to 70 years. This species, like turtles and tortoises all over the world, suffers from overexploitation. People hunt them to eat their meat and their eggs, as well as poaching them for the pet trade. Many tortoises are also sensitive to change, and can't cope when their habitats become degraded. Development on Madagascar has already destroyed more than half of the spider tortoise's habitat, leaving them homeless and vulnerable.

TURTLE TRAUMA

There are 360 known species of turtle and tortoise in the world. After surviving for more than 200 million years, sadly 127 species – more than a third of them – are today classed as endangered or critically endangered.

COELACANTH

Once known only from fossils, this fish was thought to have died out over 60 million years ago – but in 1938 it was found around the Comoro Islands, off Africa's east coast. Coelacanths live more than 100m down in the ocean, and don't survive being brought to the surface. This rare fish is unlike any other living species, and is thought to be a link to the first animals that walked on land.

ADDAX

The addax is also known as the white or screwhorn antelope. They once migrated across northern Africa in herds that could contain thousands of animals, but now, sadly, there are fewer than 300 individuals left in the wild, and they're only ever seen in groups of three or four. Local people have been hunting addaxes for centuries, and became particularly good at it during the 20th century with the invention of technologies such as high-powered rifles. Poaching by tourists and loss of habitat to the oil industry are also huge threats.

EUROPE

IBERIAN LYNX

Could this be the first cat species to become extinct in 2000 years? In 2007, it certainly looked that way – there were as few as 100 Iberian lynx surviving in the wild. That included just 25 females capable of breeding. Since then, fortunately, numbers have increased slightly, due partly to the release of animals from captive breeding programs. However, the Iberian lynx remains the world's rarest cat species, their decline linked to the overhunting of rabbits, their main prey.

THE IBERIAN LYNX WAS ONCE FOUND ACROSS SPAIN, PORTUGAL AND PARTS OF FRANCE. NOW THERE ARE ONLY TWO VERY SMALL BREEDING POPULATIONS, BOTH IN SOUTHERN SPAIN.

NORTH ATLANTIC RIGHT WHALE

Commercial whaling of the North Atlantic right whale was banned in 1935, and it has been protected ever since. Many were killed before then though, and the species is still struggling to recover. This is mostly because these animals live a long time, up to 70 years, and reproduce infrequently. Once, the species was common on both sides of the Atlantic. Now as few as 300 survive on the western side, and it's thought to be extinct on the eastern side.

DID YOU KNOW?

This species was given its common name because it was considered the "right" species for whalers to target.

MEDITERRANEAN MONK SEAL

Many thousands of Mediterranean monk seals once lived along the coastlines of Europe and north-west Africa, but after centuries of human disturbance and hunting, there are thought to be fewer than 500 left. Commercial fishing is one of the biggest threats to the species' survival, as the seals become tangled in fishing gear and drown. They are also killed by fishermen who think they compete for fish. Education campaigns for fishermen are underway to try to reduce their impact on remaining seals.

BATUECAN ROCK LIZARD

This is the rarest reptile in mainland Europe – there is just one known population, in Spain's Sierra de Francia mountain range. It only contains between 100 and 250 adult lizards. The species is threatened by the habitat impacts of road construction and tourist activity, and rare reptile collectors are also a big problem.

NEW IDENTITY

The batuecan rock lizard has only been confirmed as a separate species since 2003, and scientists still know very little about it.

EUROPEAN MINK

This mammal was a famed source of fur in the fashion industry during the past century, and overhunting for its fur seriously reduced its numbers. But that's just one in a series of threats that has brought this species to the brink. The mink is semiaquatic, and needs a clean river habitat to survive, but many of the waterways where it lives have been degraded by development and pollution. And in the 1920s, the American mink was introduced to Europe, and that has taken its toll too, the larger American species claiming the food and territory of its European cousin.

FACT!

Scientists are developing captive breeding programs to help save the European mink.

AEOLIAN WALL LIZARD

This rare lizard is found only in four small isolated populations in the Aeolian Islands, north-east of Sicily in Italy. It is feared that a disease or disaster could wipe out one of these populations, bringing the species even closer to extinction. The largest population is also thought to suffer from competition with the Italian wall lizard, an introduced species.

ENDANGERED FACT!

Captive breeding is one of the strategies being considered to ensure the survival of the Aeolian wall lizard. Introducing this species to other islands in the Aeolian chain might also give the species a better chance at survival.

ONE SOLUTION FOR THIS RARE SPIDER SPECIES MIGHT BE SPRAYING HERBICIDES TO ERADICATE THE WEED THAT HAS ALTERED ITS HABITAT.

DESERTA GRANDE WOLF SPIDER

A weed is thought to be behind the fall in numbers of the Deserta Grande wolf spider. The only place in the world this rare arachnid is known to occur is a valley on the north side of Deserta Grande Island in Portugal's Madeira archipelago. Invasion by the introduced bulbous canary-grass has drastically changed the ecosystem and altered the spider's habitat, making it critically endangered. A captive breeding program in the UK is hoping to save it from extinction.

KARPATHOS FROG

One river on the Greek island of Karpathos is the only known location for this frog. Reports from the 1960s describe the species as abundant, but it's been difficult to find since the early 1990s. These frogs rely on clean water sources that are either still or slow-running, which are now difficult to find. Like many frogs, it's likely to be very sensitive, and there may have been disturbances to its habitat that are responsible for its decline.

AMUR LEOPARD

Fewer than 35 Amur leopards are left in the wild, making this one of the world's most critically endangered big cats. Its usual habitat is the forests of north-eastern China and south-eastern Russia, although it's now thought to be extinct in China. Tragically, they have been hunted almost out of existence because poachers sell their coats for clothing and their bones for Chinese medicine.

THIS RARE LEOPARD IS PERFECTLY ADAPTED TO HUNT SMALL DEER AND HARES. IT IS FAST AND AGILE, ABLE TO RUN AT SPEEDS OF ALMOST 60KM/H, AND MAKE LEAPS 6M LONG AND 3M HIGH.

SUNDA PANGOLIN

A large international illegal trade in pangolins and their body parts threatens this mammal's survival. Their meat is eaten as a delicacy, their scales, which are made of a hardened hair-like material and cover and protect their bodies, are ground up for use in Chinese medicine, and their hide is used to make shoes. Sadly, Sunda pangolin populations that were once common in forests and grasslands across much of South-East Asia have been decimated, and they are now listed as critically endangered.

GHARIAL CROCODILE

Breeding programs have been helping this critically endangered species rebuild since the 1980s, after the population declined by 98 per cent. Eggs are collected from the wild and hatched in captivity, then once ready, the young are released back into their natural habitat. Unfortunately, settlements are growing rapidly in the areas they inhabit, so it's becoming harder to find release sites where they will be safe and thrive.

BAIJI DOLPHIN

Scientists carried out an extensive search for baiji freshwater dolphins in 2006, but none were found. Sadly, it seems this species has become the first whale, dolphin or porpoise driven to extinction by humans. The baiji's natural habitat – China's Yangtze River – is one of the world's busiest waterways, so this dolphin faced many threats, including illegal fishing, and poisoning by pesticides in wastewater discharged into the river.

SAIGA ANTELOPE

The saiga is one of the world's oldest species. Between the early 1990s and 2005, the once-healthy saiga population crashed by more than 95 per cent, from one million down to just a few thousand. This was mostly because of hunting for their meat, and the horns of the males. Overgrazing of their pastures – dry grasslands and semi-desert areas – by introduced livestock, particularly sheep, has also played a big role.

PEACOCK TARANTULA

The only place this tarantula has ever been found in the wild is a patch of land less than 100sq.km – one tenth the size of New York City – on the eastern coast of India. A large spider with a leg span of around 20cm, it lives in deep crevices in old trees. Although the home range of the spider is protected within a state forest reserve, people from nearby villages cut down trees in the reserve for firewood and timber.

This tarantula's name comes from its stunning and rare blue colouration, which unfortunately makes it popular with spider collectors worldwide.

PAINTED BATAGUR TURTLE

Overexploitation has been a major reason this Asian turtle has become critically endangered. People catch and eat the adults, and raid their nests to harvest the eggs for food. Fish farming is also a huge threat. Trading the eggs is now illegal, but on the black market they fetch more than 25 times the value of chicken eggs. The species is also popular worldwide in the pet trade because of the striking appearance of the males, which change colour during the breeding season.

MIGRATION TIME

Adults live in rivers in Borneo, Malaysia, Sumatra and Thailand, but will migrate downstream to lay their eggs in mangroves and beach nests.

PHILIPPINE EAGLE

The largest eagle in the world is found only on four islands in the Philippines. Sadly, today only 500 survive, since most of its forest habitat has been destroyed, and poisonous pesticides in the environment have reduced the number of eggs that hatch. In an effort to protect it, in 1995 it was officially declared the national bird of the Philippines.

EASTERN BLACK-CRESTED GIBBON

Over the past 45 years, this species has experienced an 80 per cent population decline, classifying it as critically endangered, and making it one of the rarest and most endangered primates in the world. The only known population of eastern black-crested gibbons in the wild today is in a tract of karst forest between Vietnam and China.

WHAT'S IN A NAME?

Also known as the Cao-vit crested gibbon, this small ape is named for the black fur on the crown of its head.

FIGHTING BACK – THE POACHING THREAT

Poaching is the biggest threat to many animal species around the world, including elephants, tigers and rhinos. Animals are poached for their body parts, or as trophies for big-game hunters. Fortunately, countries are now getting very tough with poachers. In parts of Africa and China, for example, poaching is punished with large fines and long jail sentences. Wildlife rangers are also being trained to shoot poachers who resist arrest.

MOUNTAIN PYGMY-POSSUM

This little possum lives at high altitude and loves cold weather. It's the only mammal that lives exclusively in the Australian Alps and nowhere else. It manages this by being the only marsupial that hibernates. The mountain pygmy-possum now survives in just three separate small populations.

Its habitat has suffered from the growth of Australia's ski industry and the development that's come with it. The possum is also struggling to cope with climate change. Temperatures are increasing in the Alps, and snow cover is decreasing. Feral predators, mostly cats and foxes, are another major problem that they face.

CORROBOREE FROG

This Australian alpine frog species is another of the world's mountain-dwelling amphibians that has been rapidly disappearing during the past decade. Exotic trees, including willow and birch, have invaded their native habitat, which is also being affected by climate change and disturbed by feral pigs. Additionally, huge areas have been destroyed by bushfire, and the frogs are also suffering from fatal chytridiomycosis disease.

FACT!

A captive breeding program is underway in zoos and sanctuaries in an attempt to keep this frog from extinction. It's hoped that releasing eggs and tadpoles back into home streams will help the species survive.

SUMATRAN ELEPHANT

This elephant was recognised as one of the world's most endangered animals in 2011, after half of its population disappeared in just 25 years, due to three-quarters of its habitat being destroyed in that time. Sumatra has one of the world's worst deforestation rates, further threatening this species (and others). Because of this, the surviving elephants are forced close to villages to try to find food, where they're killed when they raid crops.

Only the males have tusks and are poached, which affects breeding success.

SUMATRAN ORANGUTAN

Found on the Indonesian island of Sumatra, this is one of only three orangutan species in the world. Only 6600 survive in the wild today, mainly because of habitat destruction. The tropical forests this species calls home are being logged, mostly to make way for palm oil plantations. If this continues at current rates, these incredible orangutans are likely to become extinct within the next 50 years.

Orangutans are actually close relatives of humans, and have similar lifespans.

SUMATRAN TIGER

A century ago there were eight tiger subspecies, and about 100,000 tigers roamed Asia. Today, however, three subspecies are extinct, the South China tiger is classed as "functionally extinct" as it hasn't been seen in the wild for 25 years, and fewer than 400 Sumatran tigers survive in the wild. They are now protected, and trade in their body parts is illegal, but poaching, particularly for Chinese medicine, continues to have a huge impact.

There's a Chinese medicine market for almost every part of this tiger's body.

KAKAPO

As well as being one of the world's rarest birds, this parrot is unique in many ways. It is flightless and nocturnal, and the world's heaviest parrot, with adults reaching a weight of 3.5kg. The kakapo once lived across most of New Zealand, but it's been disappearing since colonisation began there about 800 years ago. Kakapos have no defences against predators, and introduced species such as cats, dogs, rats and stoats have played a huge role in reducing their numbers.

DID YOU KNOW?

It may be flightless, but the kakapo uses its wings to help maintain balance while it runs or climbs, and when it leaps from low trees.

BAW BAW FROG

A mid-1980s survey found more than 10,000 Baw Baw frogs living in the wild. Just 10 years later, the frog's population had crashed, declined by more than 98 per cent. Now just 250 of these tiny frogs are thought to survive, and scientists fear it will be extinct in the wild within five to 10 years. This species only lives in one very small area on the Baw Baw Plateau in Victoria, Australia.

Their habitat seems to be in good condition, but chytridiomycosis is decimating them. Fortunately Zoos Victoria is working hard to prevent their extinction.

MASSIVE LOSS

Around the world, at about the same time as the Baw Baw frog population crash, many other amphibians also disappeared. There are a few contributing factors. Most species that suffered lived at higher altitudes, so climate change has had an impact, and levels of air pollution and ultraviolet light may have become too high. The frog-killing disease chytridiomycosis, caused by the chytrid fungus, is also involved.

LARGETOOTH SAWFISH

This huge shark-like ray once lived worldwide, but it is now extinct in many places, including the United States, though some survive in northern Australia. They are overfished, because their fins are in high demand for shark fin soup, and their long tooth-studded snouts are illegally traded by collectors and used in traditional medicines. Even when these fish aren't targets, their snouts often get caught in lines and nets meant for other species, and they perish.

HAWKSBILL SEA TURTLE

Beautiful marbled markings on this sea turtle's shell have helped put the species at risk of extinction. Hawksbills are the source of real tortoiseshell, which is now illegal to trade in most countries, but has been used extensively worldwide for jewellery and ornaments.

GILBERT'S POTOROO

By the beginning of this century, the Gilbert's potoroo had become one of the world's rarest mammals – only 30 of them survived, in one isolated population in a nature reserve in Western Australia. There was real concern that it would take just one natural disaster, like a fire, for the species to disappear forever. Today, conservation efforts have brought it back from the brink, but there are still less than 100 left in the wild, so efforts continue.

WOYLIE

Native to Australia, woylies once lived right across the bottom third of the country. Now they're extinct in the east, and only a few small, isolated populations are left in the south-west. Also known as brush-tailed bettongs, they are small kangaroo-like marsupials. Their numbers crashed during the 20th century because of hunting by feral cats and foxes, and by humans for the fur trade, as well as from land clearing, and being killed by farmers as pests.

CLIMATE COST

Conservation efforts from the mid-1970s managed to increase woylie numbers, but since 2006, the total population has dropped again, thanks to habitat destruction, climate change, feral predators, and potentially from disease as well.

ORANGE-BELLIED PARROT

This parrot's green coat, distinct orange belly and unique buzzing call make it easily recognisable. They are also one of the few migratory parrots – and they only breed in Melaleuca in south-west Tasmania, making them very vulnerable. They are at high risk of becoming extinct in the next five years, with less than 50 left in the wild.

THE DEVASTATION OF OUR EXTINCTION SHAME

Australia has the worst extinction rate in the world. In the past 200 years, 34 Australian mammals became extinct – that's one-third of all mammal extinctions worldwide – along with 27 other animals and 39 plants. Feral animals such as cats, foxes and toads play a huge role – these introduced pests hunt our native species, compete for their food and shelter, destroy habitats and spread disease. Rabbits and camels are also responsible for much habitat damage – as are, sadly, humans.

SOUTHERN BENT-WING BAT

This little microbat's population has declined by two-thirds since the 1990s. Threats to the southern bent-wing bat include climate change and loss of habitat due to land grazing, pesticides, and human disturbance of roosting caves.

ENDANGERED FACT!

The bent-wing part of its name comes from its unique anatomy – the third "finger" of its wing bones is four times longer than the middle one, giving a bent appearance.

NORTH AMERICA

DID YOU KNOW?

One threat to the Hawaiian monk seal is a lack of food due to overfishing, and sharks (who are also losing resources) eating more seal pups.

HAWAIIAN MONK SEAL

There used to be three monk seal species, but the Caribbean monk seal was declared extinct in 1996. The Mediterranean species is now on the world's most endangered list, at serious risk of extinction – and if the situation doesn't improve soon for the Hawaiian monk seal, it will become extinct too. There are now only about 600 adults remaining in the wild.

ENDANGERED FACT!

Both the Caribbean and Hawaiian monk seals suffered from overhunting. During the 1800s and 1900s they were slaughtered for their meat and skin, and for the oil made from their blubber, which had many industrial uses.

ALL RUGGED UP

Sea otters have thick, water-resistant fur that keeps them warm in freezing cold ocean waters.

SEA OTTER

Sea otters are aquatic members of the weasel family. They have the densest fur of all mammals, which meant the species was hunted for their pelts to make coats and other items of clothing. As a result, the species suffered dramatic population drops during the 18th and 19th centuries. Happily, they now have legal protection throughout much of their range, to prevent extinction, but they remain endangered.

MONARCH BUTTERFLY

These striking butterflies are some of the most well known on the planet. Their orange and black wings are instantly recognisable to people living in South-East Asia, Australia, North America and parts of South America. One of the things that makes these butterflies amazing is their long migration distances, often spanning thousands of kilometres. Sadly, more than 90 per cent of North American monarchs have been wiped out over the past 20 years, due to climate change disrupting their migratory patterns, and loss of their natural habitat.

SUPER SPEEDY

They have been recorded at speeds of up to 40km/h.

RED WOLF

Red wolves are a type of wild dog that used to be common in central and eastern parts of the US. But their woodland habitat was cleared to make way for agriculture, and they were intensely hunted and trapped as pests. In 1980, they were declared extinct in the wild, and the last 17 were taken into a captive breeding program. Some have been reintroduced to the wild since, but new threats, including illegal killing, have pushed them back to the brink of extinction.

DOG DILEMMAS

Coyotes are a big threat to the reintroduced wolves. A closely related species, they inter-breed with them to produce a mixed species, and also take over their food and territory.

CALIFORNIA CONDOR

The California condor is a large vulture, which feeds by scavenging the carcasses of dead animals. Numbers fell in the 1800s after the large mammals it relied on for food were hunted to extinction. It became protected in 1967, but its numbers continued dropping, and it was extinct in the wild by 1987. A captive breeding program helped, but they are still threatened by microtrash, poaching, electrocution, and lead poisoning from the bullets they swallow when feeding on carcasses left by hunters.

WHOOPING CRANE

The whooping crane is one of the most spectacular birds native to North America. More than 1.5m tall, its white plumage and striking yellow eyes make it unmistakable. Unfortunately however, you are unlikely to see one of these birds in the wild. The species is highly endangered as a result of unregulated hunting, and the draining of wetlands for agriculture and development.

KEMP'S RIDLEY SEA TURTLE

This is one of the smallest species of turtle in the world, and the most endangered. It's also the only turtle to nest during the day, making it more vulnerable to predators. The overharvesting of its eggs, and predation by feral animals, has completely decimated it. The Kemp's ridley is now protected, but only time will tell if it will survive.

IVORY-BILLED WOODPECKER

Huge areas of this bird's forest habitat were destroyed by logging, mining and agriculture, and after last being sighted in 1945, it was declared extinct in 1996. But in 2005, it was rediscovered in the Big Woods of Arkansas in the US, leading to renewed efforts to protect this precious bird.

FLORIDA PANTHER

The Florida panther is the only subspecies of the mountain lion that remains in the eastern United States. Hunting seriously decimated the population, and it was one of the first species added to the US Endangered Species List in 1973. Today it is still endangered, with the last remaining population – of around 100 animals – found in Florida in the swamplands of the Everglades National Park and Big Cypress National Preserve. Threats include habitat loss and fragmentation and degradation caused by human development and roads.

NO ROAR

Surprisingly, not all big cats roar – and Florida panthers *can't* roar. Instead they purr, hiss, snarl, growl and yowl.

COZUMEL RACCOON

These critically endangered critters, also known as pygmy raccoons, live only on Cozumel Island, off Mexico's coast, where they forage for crabs and crayfish. The main threat to them comes from introduced animals – domestic cats and dogs attack them, compete for their food sources, and infect them with diseases and parasites. The tourist industry is also impacting them, by expanding into their habitat and creating pressure on their natural resources.

DISASTER THREAT

The pygmy raccoon population on Cozumel Island is now so small that scientists believe the whole species could be wiped out by just one hurricane.

LEMUR LEAF FROG

This little amphibian has been hit hard by chytrid – the killer fungus that's been attacking the world's amphibians since the late 1990s. The lemur leaf frog used to be abundant in the tropical forests of Costa Rica and Panama, but most of its natural populations in those two countries, as well as in Colombia, have recently either disappeared or massively declined. Sadly, habitat degradation and loss has also played a major role in this little frog's disappearance.

Since 2001, captive lemur leaf frogs have been kept at zoos in North America and Britain, where scientists are working to help them breed, and thus to save them.

GOLDEN LION TAMARIN

A golden-orange mane surrounds this monkey's face, giving the species its name. Once distributed through most of Brazil, today they are found only in a narrow strip of rainforest on the country's southern coast. Logging and habitat destruction were the greatest contributors to their decline, and by the 1970s there were fewer than 200 in the wild. Fortunately numbers have increased since then, although they're still endangered, and predators find it hard to track them because they move around so much, and never nest in the same spot more than once.

DID YOU KNOW?

There are only about 1500 individuals left in the wild. Another 500 live in zoos around the world.

BLOND CAPUCHIN

There are fewer than 200 adults of this small South American monkey still alive in the wild. They're spread between 24 separate populations along the coast, none of which are protected. The reason so few survive is because much of the blond capuchin's coastal forest home in north-east Brazil has been destroyed, mostly to make way for sugar cane farms. It is also hunted for food, and its babies are taken away as pets.

POORLY KNOWN

By the time the blond capuchin was properly described by science in 2006, it had already become one of the world's rarest primates.

BLACK-HEADED SPIDER MONKEY

These animals from Central and South America are named for the long spider-like limbs that allow them to move gracefully through the trees. Over the past 45 years, their population has suffered a 50 per cent decline due to deforestation, hunting, and being captured for the pet trade. They are easy to find due to the chatter they make while in their large groups, making them popular targets for humans and other animals.

BLUE-AND-GOLD MACAW

Macaws are parrots from Central and South America. They have beautiful feathers, which has made them popular as pets and with collectors. Sadly, overcollection has caused population crashes in many of the 17 species. Bolivia's blue-throated macaw has suffered severely from this threat. It was thought to be extinct in the wild, but a small number were rediscovered in 1992. It remains extremely rare however, making it even more prized by collectors.

BIRD BEAUTIES

The large blue-and-grey glaucous macaw hasn't been seen in the wild since the 1960s, but scientists hope it will survive out of the reach of collectors.

AMAZON RIVER DOLPHIN

Also known as the pink river dolphin, this is a shy freshwater cetacean. Numbers are declining, and it is classified as endangered due to dams that have damaged its habitat, mercury pollution from gold panning, and increased human activity on the Amazon and Orinoco rivers where it lives, which has led to several deaths.

ENDANGERED FACT!

These freshwater dolphins are born grey, then slowly turn pink as they age. And males are far pinker than females.

DARWIN'S FOX

This fox depends entirely on temperate rainforest for its survival. It was discovered by Charles Darwin in 1834 on Chiloé Island, off the coast of Chile in South America. Today most of the remaining 250 wild individuals of the species survive in a protected national park, but they are threatened by domestic dogs who attack these foxes and can also pass fatal diseases on to them.

PUFFLEG HUMMINGBIRD

These hummingbirds live in specialised forest habitats in South America's Andes Mountains, but large areas are being cleared for cattle ranches and crops, and the trees are being logged to make charcoal for use as an energy source. Education of local communities about how these activities affect pufflegs could help save them from extinction.

SOUTH AMERICAN TAPIR

Tapirs are living fossils. They've been around since the Eocene (56 to 34 million years ago), and have survived waves of extinction of other animals. They are South America's largest native land mammal, with adults reaching more than 300kg. Three different subspecies inhabit the continent – the lowland tapir, also known as the South American tapir, plus Baird's tapir and mountain tapir. They have predators in the wild, typically large cats, but their biggest threat remains human hunting and habitat destruction.

GALÁPAGOS PINK LAND IGUANA

This rare pink lizard was only discovered in 2009, and was listed as one of the world's most endangered animals by 2012. There's only one known population and it's extremely small, containing fewer than 200 adults. It lives in dry shrubland on top of an active volcano called Volcan Wolf on Isabela, the main Galápagos Island, in the tropical Pacific Ocean. Putting the iguanas further at risk, the volcano has erupted several times during the past century, most recently in 1982. It's thought that future volcanic eruptions may have a major effect on the survival of the species.

ELEGANT STUBFOOT TOAD

Most of the world's elegant stubfoot toads used to live in tropical forests in the north-west of the tiny South American country of Ecuador, but that population seems to have disappeared during the past decade. Fortunately there's a small but healthy population surviving on an island called Gorgona, off the coast of Colombia, the South American country just north of Ecuador.

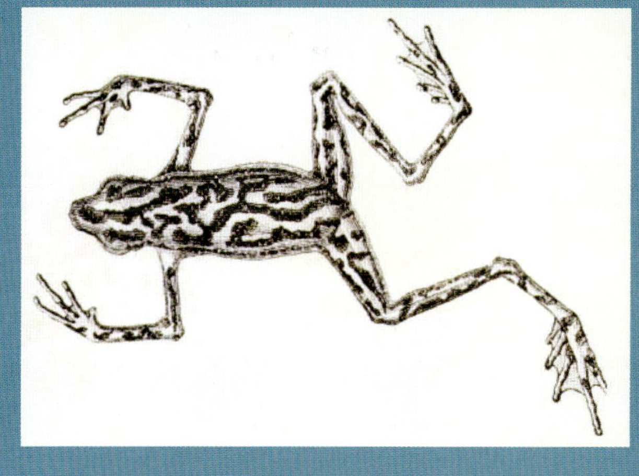

The giant whale shark is the largest extant fish known to man, reaching up to 18m in length and weighing up to 20,000kg.

EXTREME

Meet the superheroes of the animal world, beating all the records when it comes to size, speed, strength and survival!

CHIMPANZEE

Scientists continue to be astounded by the intellect of the chimpanzee – a species very closely related to humans. A chimp called Ayumu was named number one on a list of smart animals, because he outperformed humans on a memory test. Chimpanzees have also learned words, and how to use tools and play with objects.

RED-BILLED QUELEA

This small brown finch lives on Africa's savannah grasslands, and doesn't look like anything extraordinary – in fact it looks a lot like a common sparrow. But it likes hanging with other red-billed queleas so much that it can form super-colonies of up to 30 million birds! Each quelea can eat half its body weight in grain per day, which is about 10g of grain, meaning a flock of two million birds can devour up to 20 tonnes of grain in a single day. It's for this reason that, as well as being the world's most abundant bird, the red-billed quelea has another claim to fame – one of Africa's most hated birds! Its huge flocks are able to decimate entire fields of crops within hours.

EXTREME FACT!

The estimated breeding population of red-billed queleas is 1.5 billion birds.

OSTRICH

Of all the living birds, the ostrich lays the largest eggs – the biggest one ever recorded was laid in 2008 by a farmed ostrich, and weighed a massive 2.589kg.

Within a herd, all the eggs will be grouped together in one nest, then laid on by the dominant female and male. These large flightless birds can run at 70 km/h!

AFRICAN GIANT SNAIL

No other slug or snail gets bigger than this massive mollusc. The largest African giant snail ever recorded was 39.3cm long, from the end of its outstretched tail to the tip of its snout – longer than a school ruler. It weighed 900g, and had a 27.3cm-long shell. These squidgy invertebrates (distant relatives of octopuses and squids) are vegetarians, munching on plant matter, living or dead, at night in their forest habitats. From their West African home, these snails have now spread worldwide.

MASTER INVADER

The African giant snail is one of the top 100 most invasive species in the world.

CHEETAH

No large land animal moves faster than this big cat. The main secret to the cheetah's speed is its long and flexible spine, which helps it cover almost 7m with every stride. Another trait that sets cheetahs apart from other fast four-limbed movers is that they can increase the rate of strides they take as they speed up. Other fast creatures, such as race horses, can't do this – they take the same number of strides per second no matter what speed they reach. The light, slender build of cheetahs also helps these big cats move fast in short bursts, as do their sharp claws, which grip the ground to provide traction as they run.

KLIPSPRINGER

These African antelopes are just 45 to 60cm tall at the shoulder – but can jump up to 10 times their height, making them the highest jumpers of all mammals relative to body size. This extraordinary capability makes it hard for potential predators such as leopards, hyenas and baboons to grab them. They walk daintily on the tips of their hooves, as though on tippy-toes.

SPRINGBOK

When these antelopes were migrating across the western plains of southern Africa, they formed the largest known herds of any animals on earth. These huge aggregations could be 145km long and contain tens of millions of springboks. In the past 120 years though, the human impacts of hunting, urbanisation, farming and development have severely impacted their numbers and prevented their migration.

One of the main reasons animals form herds is so they can work together to keep watch and warn each other about any approaching predators.

AFRICAN BUSH ELEPHANT

When it comes to records for the heaviest living land animals, elephants top the list. The African bush elephant comes first, with males reaching weights of up to 7 tonnes. African forest elephants and Asian elephants can reach more than 5 tonnes each, enough to place them second and third heaviest. Elephants are vegetarians, eating mostly roots, grass, fruits and bark, and have to eat around 130kg a day to maintain their weight.

Despite their huge size, African bush elephants can run at up to 40km/h, and maybe even faster!

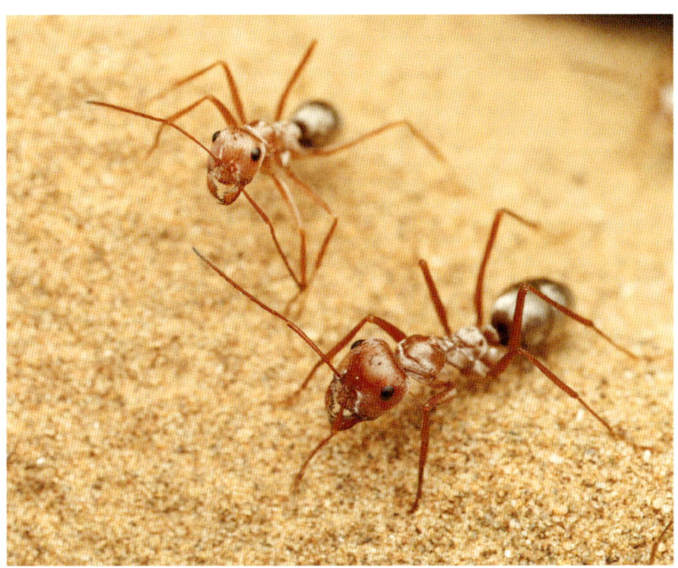

SAHARA DESERT ANT

This tiny insect is one of the most heat-tolerant creatures in the world, and one of the few able to withstand the searing midday heat of the Sahara Desert, the hottest terrestrial habitat on the planet. It can leave its sandy burrow at the hottest time of the day, when the surface temperature reaches up to 70°C, but it must keep moving to maintain its body below the critical maximum temperature of 55.1°C.

FAST FORAGERS

Sahara Desert ants brave incredibly hot conditions to scavenge for the carcasses of invertebrates such as spiders.

GOLIATH FROG

Most of the world's frog species are small enough to fit neatly on the palm of a human hand, but that's certainly not the case for the goliath frog. It can reach the proportions of a newborn human baby – longer than 30cm, and more than 3kg in weight!

EXTREME FACT!

Goliath frogs have no vocal sac and so can't call for mates as other frogs do.

EUROPE

SPERM WHALE

The animal kingdom is full of squawking, screeching, rumbling loudmouths, but the loudest of them all is the sperm whale. These huge deep-ocean creatures communicate by producing clicking noises, the loudest of which have been measured at 230 decibels (dB). To give an idea of how loud that is, a jet plane taking off is about 165dB. Sperm whale clicks are quick – no more than 30 milliseconds – but they can be heard by another sperm whale more than 16km away.

THE SPERM WHALE HAS ANOTHER EXTREME FEATURE – IT HAS AN 8KG BRAIN, THE LARGEST OF ANY ANIMAL THAT HAS EVER LIVED.

LONG-FINNED PILOT WHALE

The neocortex is a part of the brain that is particularly well-developed in humans. Neuroscientists always assumed we have more nerve cells in the neocortex than any other species – so they were shocked when a 2014 study of the brains of long-finned pilot whales in the North Atlantic found that this species has more brain cells in the neocortex than any other species ever studied – and almost twice as many neurons in this part of the brain as humans. They are still wrestling with what this might mean about the intelligence of pilot whales.

GIANT SQUID

The giant squid is the largest invertebrate alive today, but no one is sure just how big it gets. Suspected to be the cause of many sea monster myths, this enormous mollusc species is rarely seen, as it lives at great depths in the Atlantic Ocean, feeding on deep-sea fishes and squid. But there have been reports of 18m-long specimens that have washed up on beaches.

HAGFISH

Hagfish are ancient fish with four hearts, part of a primitive system that circulates blood around their bodies. One heart acts as the main pump, while the other three help out. Another extreme hagfish feature is how they eat – they burrow deep into the dead bodies of other animals to eat their rotting flesh from the inside out. Then there's their response to predators. When something tries to grab them, they produce such huge quantities of slime that they're impossible to get or keep hold of.

SPRINGTAIL

Ever feel you're not alone? Chances are you never have been. The world is crawling with life, and we're not just talking bacteria or other microscopic flora and fauna. Among the most abundant creatures in the world are the insect-like springtails. With most between 2–6mm long, many different types are visible to the naked eye, if you look out for them. Springtails live in high humidity environments, where they feed on substances such as mould, bacteria and decaying plant matter. Scientists estimate their numbers to be as high as 100,000 per cubic metre, even in places where they are living side-by-side with people.

BOOTLACE WORM

The bootlace worm, a marine invertebrate found in sediment along the British coast, could make the cut as the world's longest animal. Specimens are often found at between 5–15m long, but they can reach much greater lengths, with reports of individuals that were able to be stretched out to 55m, longer than any other creature – even the blue whale!

ARCTIC FOX

The small Arctic fox is the only member of the dog family that changes colour as the weather changes. It has a thick white coat when its habitat is shrouded in snow and it needs to be camouflaged so it can sneak up on prey – and not be seen by predators. But as the snow and ice melts in spring and summer, and the Arctic tundra is transformed into warmer tones, the winter coat moults and a new lighter coat grows, a two-tone tawny brown in colour to match the new landscape.

ARCTIC TERN

These small birds embark on a huge annual migration that takes them around the planet and back, further than any other bird. Every year they fly from their breeding grounds in the Arctic to the other end of the world, to spend the Southern Hemisphere summer feeding on small fish and crustaceans in the waters off Antarctica, then they fly back again – a return journey of more than 70,000km! They're long-lived birds, and many will complete this return trip more than 20 times.

NORTHERN FULMAR

Fulmar chicks are as fluffy and cute as any other baby bird, but they're certainly not as defenceless. Their response to a potential threat or predator is to vomit up a vile orange concoction from deep in their gut, which they can shoot up to 1.8m. This acts a deterrent not only because it smells disgustingly of rotten fish, but also because it is so sticky that it's almost impossible to remove, even with repeated scrubbing.

CUVIER'S BEAKED WHALE

No other mammal dives deeper than a Cuvier's beaked whale. Using satellite tags to track whales off the Californian coast, scientists have recorded adults of this species diving to depths of almost 3km. Exactly how they manage to withstand the immense pressure at these depths is not yet understood, but they're thought to have collapsible rib cages to reduce air pockets that would make them buoyant as they dived.

GREENLAND SHARK

These incredible creatures are massive – weighing in at more than 1000kg and measuring over 7m in length. Found in the icy waters of the Arctic Ocean and the North Atlantic, these sharks are also incredibly long-lived. Researchers recently identified one that is 270 years old, and they believe there are some still older yet that will be found. What is even more astonishing is that they only start breeding when they are 150 years old.

IMMORTAL JELLYFISH

These jellyfish are tiny, just the size of a fingernail, but that is not what is the most amazing thing about them. This jellyfish can do something that humans have been unable to do – they can revert back to being young again, literally. When the immortal jellyfish reaches adulthood, if facing starvation or other hardship, they can revert back to being a baby. Depending on where they live, they can also have eight or more arms.

ASIA

GIANT PANDA

Although giant pandas have a vegetarian diet, eating primarily bamboo, as members of the bear family they still have the digestive system of a carnivore, a meat-eating animal. But as they've chosen a bamboo-chomping lifestyle (believed to be because it's abundant and they don't have to fight for it), pandas have evolved the biggest molar teeth of any carnivore, which, along with their strong jaws, allow them to crush hard bamboo stems so they can digest them along with the leaves.

GIANT HUNTSMAN SPIDER

The famous goliath birdeater might be the largest spider when measured by weight, but the giant huntsman takes the prize when it comes to sheer size. This spider was only discovered in 2001, in a cave in the South-East Asian country of Laos, which lies north of Thailand. If this slender spider species sat on a dinner plate, its legs would stretch right across. No other spider has such a long leg span – up to 30cm.

GIANT CLAM

This creature is extreme in a number of ways, but perhaps most bizarre are its eyes – it's thought to have more than any other animal. The uppermost exposed part of a giant clam's flesh is covered with them, and a fully-grown specimen can literally have hundreds of eyes. And while these might not be the sort of complex eyes humans have, they can detect movement as well as light and dark. On top of all that, it's also the largest bivalve mollusc on the planet.

BUMBLEBEE BAT

The tiny Kitti's hog-nosed bat grows no longer than 2–3cm, and weighs just 2g or less! It's the same size as a bee, hence its common name. It's the smallest mammal in the world, and lives in caves in Thailand and Myanmar. It survives on small insects that it catches using echolocation.

CHINESE GIANT SALAMANDER

This huge relative of frogs and toads can grow to a length of almost 1.8m. But sadly there are very few left now, and they are rarely seen at that size.

EXTINCTION THREAT

This salamander is critically endangered, due partly to the use of its parts in traditional Chinese medicine.

BABIRUSA

It's known that the tusks of elephants and rhinoceroses and the antlers of deers and antelopes are weapons, used against predators and to spar against males of the same species. But no one is quite sure about the purpose of the tusks on a male babirusa, also known as a deer-pig, although they are thought to potentially protect their eyes. These extraordinary features are actually canine teeth that grow out from the jaw and up through the nose. The lower canines of a male babirusa are also sharp and protruding. There are four babirusa species, all of them found in Indonesia.

BABIRUSA TUSKS ARE SO BRITTLE THAT THEY BREAK OFF EASILY, MAKING THEM USELESS AS WEAPONS OR FOR ANY OTHER PURPOSE OTHER TUSKS ARE USED FOR.

ASIAN ELEPHANT

All elephant species have an exceptionally long gestation period (the full length of an average pregnancy). Elephant mums are pregnant for up to 23 months, which is longer than any other mammal, and more than twice as long as the human gestation period of nine months. Asian elephants are pregnant for around 620 days – and their African relatives for up to 655 days. Fortunately they then have a two-year break.

JAPANESE SPIDER CRAB

When it comes to width, no crab is bigger than the Japanese spider crab, which is found mostly in waters up to 600m deep in the Pacific Ocean off the coast of Japan. Although the carapace – the shell covering the main body – of these crabs is usually no longer than about 37cm, the leg span of adults can be massive – as much as 4m, which is wider than a bus! And it can weigh up to 20kg.

ZEBRAFISH

Scientists are looking to the zebrafish to learn how to fix human hearts, because they have a special capability for regeneration. When a zebrafish heart is wounded, the site is sealed off by a clot, and the surrounding heart muscle grows over it. Soon the heart is almost as good as new, and pumping as strongly as ever.

PEACOCK MANTIS SHRIMP

This stunning underwater creature, also known as the rainbow mantis shrimp, has two extreme claims to fame. Firstly, this colourful crustacean has a dark side – it's a voracious predator that likes to punch, and it does so at a startling 80km/h – faster than has been recorded for any other animal. It even has two front claws specially adapted for punching through the tough shells of molluscs to reach the soft tissue inside, so it can eat it. The other astounding feature of peacock mantis shrimp is their incredible eyesight. They have highly specialised eyes that can see colour better than most animals, including us. Even more impressive is that they can also detect several different types of light, including infrared and ultraviolet, and even a type of polarised light no other animal is known to be able to detect.

BLACK MARLIN

Like many animal records, not everyone agrees on the species that takes the title of fastest fish. The Indo-Pacific sailfish is regularly at the top of this list, as it's been recorded slicing through the tropical waters of the Indian and Pacific oceans at more than 110km/h in short bursts. More recently however, another fish has laid claim to the title. Based on the rate at which hooked black marlins unwind fishing line from a reel, scientists have estimated that this fish can reach a top speed of 129km/h!

RETICULATED PYTHON

South-East Asia's reticulated python doesn't grow to be as heavy as South America's anaconda, but it certainly seems to grow longer. There have been many reports of individual pythons that are longer than 6m, and in 2016 a python was caught on the Malaysian island of Penang that was estimated to be 8m long, which weighed in at a terrifying 250kg.

EMPEROR PENGUIN

This bird breeds in the coldest place on earth – Antarctica in winter. There, on the open ice, male emperor penguins huddle together, each with a single egg held up off the ground on their feet, and kept warm in a feathered belly pouch. Standing like this, they'll resolutely face months of the worst winter conditions that occur anywhere, including wind chills as low as -60°C, and 200km/h blizzards. Male emperor penguins won't feed during that entire time either, not until their partner returns from foraging to relieve them of their egg-incubating duties.

EMPERORS, THE LARGEST OF THE WORLD'S 18 PENGUIN SPECIES, REACH A HEIGHT OF 1.15M AND UP TO 45KG IN WEIGHT.

LION'S MANE JELLYFISH

This is not only the longest jellyfish known to science, but also a contender for the longest animal of all. The biggest known specimen was washed up on a beach, and had a 2.3m-wide bell and 37m-long tentacles, exceeding the 30m length reached by blue whales, but not quite as long as Europe's bootlace worm.

CENTRAL BEARDED DRAGON

When it comes to the world's fastest lizard, there are at least two contenders to consider. One is the black spiny-tailed iguana of South and Central America, which has officially been clocked running at almost 35km/h. But the other is Australia's central bearded dragon, a 60cm-long creature that rises up on its hind legs when threatened by a predator and races across the landscape at what's estimated to be around 40km/h – which could make it even more worthy of the title of world's fastest lizard.

TUATARA

Despite their appearance, these native New Zealand creatures are not actually lizards. They are the only surviving members of the Rhynchocephalia family, an ancient group that flourished as long as 200 million years ago. All but the tuatara became extinct around 60 million years ago. Tuataras have a third eye, known as a parietal eye. It's located on the top of the head, although by the time the tuatara reaches four months old, it's usually covered in skin and scales. It isn't used for vision, but for regulating temperature and sleep.

QUEEN ALEXANDRA'S BIRDWING

The largest butterflies in the world are female Queen Alexandra's birdwings, which have a wingspan of almost 30cm. The species is sexually dimorphic, meaning that males and females look very different to each other – the males are smaller than the females, and have such different physical markings that they look like a separate species. These stunning insects are found only in coastal rainforests in eastern Papua New Guinea, and are endangered due to destruction of their habitat for palm oil plantations.

BLUE WHALE

It's thought that no other animal that's ever lived has achieved the size of the blue whale, not even the biggest dinosaurs. The largest blue whale on record weighed almost 200 tonnes, and was more than 30m long. But while they might be the biggest creatures, blue whales prey only on tiny crustaceans called krill, strained from polar waters. They eat up to 4 tonnes of krill a day!

BULL ANT

There are about 90 species in the Australian ant genus *Myrmecia*, most commonly referred to as bull ants, and some of these are potentially harmful to humans. Bull ants have caused more human deaths than any other ant – at least six people in Australia have died from anaphylaxis, an extreme allergic reaction, after being bitten. Five of these occurred on the Australian island of Tasmania, where a study found that up to three per cent of the population is allergic to their venom. That's twice the level of people allergic to the venom of European honey bees.

PEREGRINE FALCON

During its hunting dives, the peregrine falcon can reach more than 320km/h as it plummets from great heights to swoop in and grab or stun prey. These streamlined, powerfully built birds of prey hunt other birds, as well as mammals such as rabbits. When chasing other birds, peregrine falcons can also fly at great speeds, although not quite as fast as they achieve when plummeting under gravity.

The oldest peregrine ever recorded was 19 years and 9 months old in the United States.

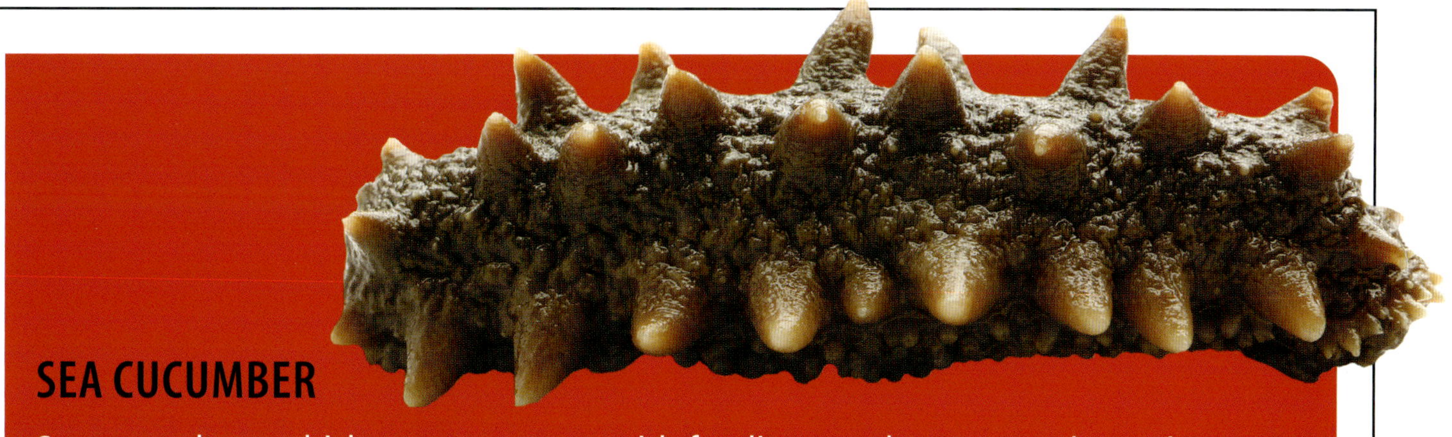

SEA CUCUMBER

Sea cucumbers, which are relatives of starfish and sea urchins, and not vegetables at all, have one of the most extreme forms of defence against would-be predators. When threatened, these cylinder-shaped creatures with feeding tentacles and tube feet, can expel their internal organs, particularly their guts, sending them out through either their mouth or the other end, depending on the species. It's a behaviour known as evisceration, and is used to scare off the crabs and fish that prey on them. The organs regenerate inside the sea cucumbers within five weeks, and they appear to suffer no long-term ill effects from doing this.

TASMANIAN GIANT CRAB

This crustacean is found in the cold ocean waters off southern Australia. Males grow to weights of more than 17kg, making it the world's second-heaviest crab after the Japanese spider crab. The shells of the largest Tasmanian giant crabs are up to 46cm wide.

KIWI

The kiwi, a flightless bird found only on the Pacific Island nation of New Zealand, produces the largest egg of any bird in relation to the mother's body size. These eggs are six times bigger than would be expected for a bird the size of a kiwi, and take up about 25 per cent of the mother's body. While the ostrich produces the largest egg of any bird, it only takes up about two per cent of its mother's body.

NORTH AMERICA

STRIPED SKUNK

There are a few animals that use smell as a form of self-defence, and skunks are perhaps the best known. These mammals don't even try to outrun a threat – instead they simply turn their backside to their attacker and spray a noxious substance from large anal glands near the base of their tail. They can spray this oily, hard-to-remove substance out across a distance of 3m, and most predators run away rather than taking on a skunk.

BANANA SLUG

Slugs and snails have reputations as slow movers… extremely slow movers! And banana slugs seem to be among the slowest. While the average garden snail moves at a rate of about 78cm per minute, these bright yellow North American slugs move at, well, considerably less than a snail's pace – at about 17cm per minute.

REINDEER

The eyes of most mammals can only detect light in the visible part of the spectrum, but reindeers have a unique visual adaptation that means they are the only mammals that can also see ultraviolet (UV) light. This is believed to be an adaptation to the Arctic environment where they live, as snow and ice reflect a lot of UV light. In contrast lichen, which is eaten by reindeers, and the bodies of their predators, all absorb UV light, so being able to see this light makes it much easier for reindeers to locate their food and avoid predators such as wolves.

MAYFLY

All mayflies have short adult lives, but females take it to the extreme, living as adults for only five minutes – just enough time to mate and lay eggs. After hatching, the juveniles, known as nymphs, live in watery habitats, feeding on algae or insect larvae for up to two years, before emerging as winged adults. Then, within a few minutes, the next generation is created and the life cycle begins again.

AMERICAN PYGMY SHREW

This is North America's smallest mammal by weight. Adults grow to a maximum of just 4g, and maintaining body heat at this tiny size is a constant challenge. It's why these shrews have one of the fastest heartbeats of any mammal – 1200 beats per minute. To sustain this, they need to eat three times their body weight every day, so their life is spent almost constantly foraging for food such as beetles, spiders and other small invertebrates, and they can only ever sleep for a few minutes at a time.

PERIODICAL CICADA

Nearly all cicadas spend most of their life underground as juveniles, before emerging above to experience a short life as adults. Periodical cicadas are remarkable for their synchronicity – each generation emerges at once – and for their lifespan. They spend either 13 or 17 years underground as nymphs, then live for just four to six weeks as adults, breeding, laying eggs, then dying, their purpose fulfilled.

BRISTLEMOUTH FISH

These tiny bioluminescent fish don't just occur in the waters off North America, but in deep water throughout the world's temperate ocean habitats. Scientists now believe that bristlemouths, also known as lightfishes, are the most abundant fish of all. There are thought to be more bristlemouths than any other vertebrate – hundreds of trillions of them, and possibly even quadrillions, which is thousands of trillions – far too many to get your head around!

AMERICAN WOODCOCK

American woodcocks fly at a very respectable rate of up to 45km/h when they migrate – but during an elaborate aerial courtship dance performed by males to attract mates, they move so slowly that they almost stall in mid-air, dropping to just 8km/h, the slowest flying speed recorded for a bird. Each male has a few preferred sites where he performs his courtship flights, and also sings to impress prospective mates.

ARCTIC GROUND SQUIRREL

Alaska's Arctic ground squirrels have adapted to allow them to survive their harsh environment. They hibernate for up to eight months a year, and researchers have discovered that during this long sleep they drop their body temperature to below freezing (-2.9°C) – the lowest ever recorded for a mammal. Their major organs slow, and others shut down, slowing the metabolism so they can survive their long hibernation, when they don't eat or drink anything.

MASSIVE LOSS

The Arctic ground squirrel loses almost half of its body weight during the eight months it hibernates, so before it sleeps it bulks up to double its size.

WOOD FROG

Very few animals can be frozen and not die, but the wood frog is one that can. During the long Alaskan winters it buries itself in the ground and enters a deep state of hibernation. It doesn't breathe, its heartbeat stops, and most of the water in its body turns to ice. At the height of winter, its body temperature hovers between -10°C and -60°C, then as spring arrives and it thaws out, bodily functions return and it hops off to begin breeding as soon as possible.

TEXAS LONGHORN

No prizes for guessing how this animal got its name! The Texas longhorn is a North American cattle breed, and the average width of the horns on the bulls is a massive 1.8m from tip to tip – the length of a tall human. It can be even longer in some steers and extraordinary cows, but there's one particular steer who makes the average horn length seem short. In 2019, Poncho Via broke the Guinness World Record with a horn spread that is a massive 3.236m wide.

SOUTH AMERICA

TOCO TOUCAN

The toco is the largest toucan species, and it has the biggest beak for its body size of any bird – it's one-twentieth of the toucan's total weight, and one-third of its total length. Bright but clever colouration ensures that these birds are perfectly camouflaged in their homes in Brazil's forest and savannahs, but then they go and blow their cover with some of the most varied and noisy vocalisations of any bird. They produce a persistent rattle that sounds like they are having an ongoing conversation, and they create deep croaks that they also repeat incessantly.

DID YOU KNOW?

Toco toucans are mostly frugivores – they primarily eat fruits that they pick from the forest canopy.

DARWIN'S FROG

There is a range of ways animal fathers get involved in the care of their offspring, but the strategy taken by male Darwin's frogs is one of the most bizarre. After the female lays her eggs in leaf litter, the male fertilises them and guards them for up to four weeks until tadpoles can be seen wriggling inside the eggs. The male then ingests the eggs and holds them in his vocal sac. He keeps the developing tadpoles in his mouth until they change into tiny frogs and hop out of his mouth.

HERCULES BEETLE

As its common name suggests, this small creature, which is large for an insect, has extraordinary strength. It's found in the jungles of South America, and is famously able to lift objects 850 times its own body weight. That would be like a human weight lifter hoisting 65 tonnes above their head! The other weird feature of these beetles is the enormous pincers that protrude from the heads of males. They look like massive horns, which is why the group this insect belongs to is known as the rhinoceros beetles.

BIG BOYS

Hercules beetles are not only one of the strongest beetles, but also one of the largest insects – adult males can grow up to 17cm.

TUBE-LIPPED NECTAR BAT

The 8.5cm-long tongue of this bat stretches one-and-a-half-times its body length, which makes it longer than the tongue of any other mammal when compared to their overall body size. It's only beaten by the ridiculously long tongues of chameleons, a group of lizards found mostly on the African island of Madagascar.

GREEN ANACONDA

Two species vie for the title of biggest snake. The reticulated python is the longest, but when it comes to sheer bulk, they don't get any heavier than a fully grown green anaconda. The largest anaconda to be verified weighed 97.2kg and was 5.2m long. There are, however, reports of a specimen that was more than 8.5m long, measured 30cm around its belly, and was estimated to weigh more than 220kg!

BEE HUMMINGBIRD

Found only in Cuba, this beautiful creature is the world's smallest bird. They are so tiny – just 5.5cm long when fully grown – that they are often mistaken for bees, hence their name. And each one only weighs about 2g! Their precious miniature wings beat at 80 times a second during flight, and when they are courting that can increase to a whopping 200 times.

BLACK PIRANHA

For its body size, the black piranha has the most powerful bite of any carnivorous fish. It achieves this by having jaw muscles of an extraordinary size for its overall body weight, as well as having a highly specialised jaw-closing mechanism. This gives it a bite force equivalent to 30 times its body weight, which is three times more than the bite force that could be exerted by an American alligator if it was the same size.

EXTREME FACT!

Piranhas have razor sharp teeth similar to sharks, and will lose and replace all their teeth multiple times during their lifetime. The word piranha means fish tooth in the Brazilian Tupi language.

GALÁPAGOS TORTOISE

This is the largest living tortoise species, and one of the slowest moving animals for its size. It can usually only make a top speed of 1.6km/h, and often moves more slowly, but males can be more enthusiastic during the mating season, when they've been tracked travelling as far as 13km in just two days on treks in search of females. Galápagos tortoises can reach a weight of more than 250kg, and have a shell that's up to 1.5m long.

VERY LONG LIVES

Galápagos tortoises are among the longest living vertebrates, with a lifespan of 150 years, perhaps helped by the fact they rest for 16 hours a day!

HOATZIN

South America's swamps seem an appropriate home for the bizarre bird that's widely considered to be the stinkiest – the hoatzin, also known as the stinkbird. It appears to be a relic species, with many peculiar features, and scientists are baffled about which other living species are its closest relatives. It's a herbivore that uses bacteria in part of its gut to ferment and help digest the fruit and leaves it eats. Although cattle do a similar thing, it's not seen in any other birds.

Recent research suggests the hoatzin is the only surviving member of a bird group that branched off on its own just after the extinction of the dinosaurs.

LARGE HAIRY ARMADILLO

When it comes to sleepy animals, this one is up there with the best. Like other big sleepers, such as Australia's koalas and South America's sloths, they survive on food that's very low in nutritional value, so they don't have the energy to do much else but eat and sleep. Nocturnal, they forage for beetles and other insects, then spend the rest of their time – up to 20 hours a day – in their burrow, not necessarily always asleep, but certainly hard at rest.

SPECIES INDEX

First published in 2021.
Australian Geographic
52–54 Turner St, Redfern NSW 2016
02 9136 7206
editorial@ausgeo.com.au
australiangeographic.com.au

Authors: Liz Ginis, Karen McGhee and Pete Tuskan
Creative director: Mike Ellott
Designer: Mel Tiyce
Editor: Martine Allars
Chief Sub-Editor: Serene Conneeley
Photos: Australian Geographic, Getty Images, Shutterstock, Chris Lukhaup/CC-Wikipedia (panda ant) and Mark Carwardine (baiji dolphin)
Production: Andy Franks
Commercial and Rights Manager: Simone Aquilina

Australian Geographic
Managing Director: Jo Runciman
Editor-in-Chief: Chrissie Goldrick

Funds from the sale of this book go to support the Australian Geographic Society, a not-for-profit organisation dedicated to sponsoring conservation and scientific projects, as well as adventures and expeditions.

Printed by 1010 Printing Asia Limited.

A catalogue record for this book is available from the National Library of Australia